Edward L. Wolf

Nanophysics of Solar and Renewable Energy

Related Titles

Wengenmayr, R., Bührke, T. (eds.)

Renewable Energy

Sustainable Energy Concepts for the Future

2012

ISBN: 978-3-527-41108-5

Vogel, W., Kalb, H.

Large-Scale Solar Thermal Power

Technologies, Costs and Development

2010

ISBN: 978-3-527-40515-2

Quaschning, V.

Renewable Energy and Climate Change

2010

ISBN: 978-0-470-74707-0

Wolf, E. L.

Quantum Nanoelectronics

An Introduction to Electronic Nanotechnology and Quantum Computing

2009

ISBN: 978-3-527-40749-1

Würfel, P.

Physics of Solar Cells

From Basic Principles to Advanced Concepts

2009

ISBN: 978-3-527-40857-3

Freris, L., Infield, D.

Renewable Energy in Power Systems

2008

ISBN: 978-0-470-01749-4

De Vos, A.

Thermodynamics of Solar Energy Conversion

2008

ISBN: 978-3-527-40841-2

Wolf, E. L.

Nanophysics and Nanotechnology

An Introduction to Modern Concepts in Nanoscience

2006

ISBN: 978-3-527-40651-7

Edward L. Wolf

Nanophysics of Solar and Renewable Energy

WILEY-
VCH

WILEY-VCH Verlag GmbH & Co. KGaA

The Author

Prof. Edward L. Wolf
Polytechnic Institute of the New York University
Brooklyn, USA
email: ewolf@poly.edu

Cover picture
Pictures clockwise:
The sun
photographed by NASA's SOHO spacecraft
© NASA 2004

The flexible solar module
(Credit: Copyright Fraunhofer ISE)

Solar panels
Part of the Solar Farm at PT.LEN Industri, Indonesia's largest solar cell producer and importer. This 900 square meter farm generates enough electricity to power their solar factory and the employee's cafetaria.
Photograph by Chandra Marsono, 2008

Pillared graphene consists of CNTs and graphene sheets combined to form a 3D network nanostructure
© SPIE 2009
George Dimitrakakis, Emmanuel Tylianakis, and George Froudakis. Designing novel carbon nanostructures for hydrogen storage. SPIE Newsroom doi 10.1117/2.1200902.1451

All books published by **Wiley-VCH** are carefully produced. Nevertheless, authors, editors, and publisher do not warrant the information contained in these books, including this book, to be free of errors. Readers are advised to keep in mind that statements, data, illustrations, procedural details or other items may inadvertently be inaccurate.

Library of Congress Card No.: applied for

British Library Cataloguing-in-Publication Data
A catalogue record for this book is available from the British Library.

Bibliographic information published by the Deutsche Nationalbibliothek
The Deutsche Nationalbibliothek lists this publication in the Deutsche Nationalbibliografie; detailed bibliographic data are available on the Internet at http://dnb.d-nb.de.

© 2012 Wiley-VCH Verlag & Co. KGaA, Boschstr. 12, 69469 Weinheim, Germany

All rights reserved (including those of translation into other languages). No part of this book may be reproduced in any form – by photoprinting, microfilm, or any other means – nor transmitted or translated into a machine language without written permission from the publishers. Registered names, trademarks, etc. used in this book, even when not specifically marked as such, are not to be considered unprotected by law.

Composition   Thomson Digital, Noida, India

Printing and Binding   Markono Print Media Pte Ltd, Singapore

Cover Design   Schulz Grafik-Design, Fußgönheim

Print ISBN: 978-3-527-41052-1 (HC)
              978-3-527-41046-0 (SC)
ePDF ISBN: 978-3-527-64631-9
ePub ISBN: 978-3-527-64630-2
mobi ISBN: 978-3-527-64629-6
oBook ISBN: 978-3-527-64628-9

Printed in Singapore
Printed on acid-free paper

In Memory of Ned

Edward O'Brien Wolf

1973–2011

Contents

Preface XIII

1	**A Survey of Long-Term Energy Resources** 1	
1.1	Introduction 1	
1.1.1	Direct Solar Influx 6	
1.1.1.1	Properties of the Sun 6	
1.1.1.2	An Introduction to Fusion Reactions on the Sun 10	
1.1.1.3	Distribution of Solar Influx for Conversion 13	
1.1.2	Secondary Solar-Driven Sources 14	
1.1.2.1	Flow Energy 14	
1.1.2.2	Hydroelectric Power 18	
1.1.2.3	Ocean Waves 20	
1.1.3	Earth-Based Long-Term Energy Resources 22	
1.1.3.1	Lunar Ocean Tidal Motion 22	
1.1.3.2	Geothermal Energy 24	
1.1.3.3	The Earth's Deuterium and its Potential 25	
1.1.4	Plan of This Book 26	
2	**Physics of Nuclear Fusion: the Source of all Solar-Related Energy** 27	
2.1	Introduction: Protons in the Sun's Core 28	
2.2	Schrodinger's Equation for the Motion of Particles 30	
2.2.1	Time-Dependent Equation 32	
2.2.2	Time-Independent Equation 32	
2.2.3	Bound States Inside a One-Dimensional Potential Well, $E > 0$ 33	
2.3	Protons and Neutrons and Their Binding 35	
2.4	Gamow's Tunneling Model Applied to Fusion in the Sun's Core 35	
2.5	A Survey of Nuclear Properties 43	

3 Atoms, Molecules, and Semiconductor Devices 49

- 3.1 Bohr's Model of the Hydrogen Atom 49
- 3.2 Charge Motion in Periodic Potential 52
- 3.3 Energy Bands and Gaps 53
- 3.3.1 Properties of a Metal: Electrons in an Empty Box (I) 57
- 3.4 Atoms, Molecules, and the Covalent Bond 60
- 3.4.1 Properties of a Metal: Electrons in an Empty Box (II) 66
- 3.4.2 Hydrogen Molecule Ion H_2^+ 69
- 3.5 Tetrahedral Bonding in Silicon and Related Semiconductors 71
- 3.5.1 Connection with Directed or Covalent Bonds 72
- 3.5.2 Bond Angle 72
- 3.6 Donor and Acceptor Impurities; Charge Concentrations 73
- 3.6.1 Hydrogenic Donors and Excitons in Semiconductors, Direct and Indirect Bandgaps 75
- 3.6.2 Carrier Concentrations in Semiconductors 76
- 3.6.3 The Degenerate Metallic Semiconductor 79
- 3.7 The PN Junction, Diode I–V Characteristic, Photovoltaic Cell 80
- 3.8 Metals and Plasmas 84

4 Terrestrial Approaches to Fusion Energy 87

- 4.1 Deuterium Fusion Demonstration Based on Field Ionization 88
- 4.1.1 Electric Field Ionization of Deuterium (Hydrogen) 94
- 4.2 Deuterium Fusion Demonstration Based on Muonic Hydrogen 96
- 4.2.1 Catalysis of DD Fusion by Mu Mesons 101
- 4.3 Deuterium Fusion Demonstration in Larger Scale Plasma Reactors 102
- 4.3.1 Electrical Heating of the Plasma 103
- 4.3.2 Scaling the Fusion Power Density from that in the Sun 104
- 4.3.3 Adapt DD Plasma Analysis to DT Plasma as in ITER 104
- 4.3.4 Summary, a Correction, and Further Comments 110

5 Introduction to Solar Energy Conversion 115

- 5.1 Sun as an Energy Source, Spectrum on Earth 115
- 5.2 Heat Engines and Thermodynamics, Carnot Efficiency 117
- 5.3 Solar Thermal Electric Power 119
- 5.4 Generations of Photovoltaic Solar Cells 122
- 5.5 Utilizing Solar Power with Photovoltaics: the Rooftops of New York versus Space Satellites 125
- 5.6 The Possibility of Space-Based Solar Power 126

6 Solar Cells Based on Single PN Junctions 133

- 6.1 Single-Junction Cells 133
- 6.1.1 Silicon Crystalline Cells 136
- 6.1.2 GaAs Epitaxially Grown Solar Cells 141
- 6.1.3 Single-Junction Limiting Conversion Efficiency 141

6.2	Thin-Film Solar Cells versus Crystalline Cells	145
6.3	CIGS ($CuIn_{1-x}Ga_xSe_2$) Thin-Film Solar Cells	147
6.3.1	Printing Cells onto Large-Area Flexible Substrates	147
6.4	CdTe Thin-Film Cells	151
6.5	Dye-Sensitized Solar Cells	153
6.5.1	Principle of Dye Sensitization to Extend Spectral Range to the Red	154
6.5.2	Questions of Efficiency	155
6.6	Polymer Organic Solar Cells	155
6.6.1	A Basic Semiconducting Polymer Solar Cell	156
7	**Multijunction and Energy Concentrating Solar Cells**	**157**
7.1	Tandem Cells, Premium and Low Cost	158
7.1.1	GaAs-based Tandem Single-Crystal Cells, a Near Text-Book Example	158
7.1.2	A Smaller Scale Concentrator Technology Built on Multijunction Cells	162
7.1.3	Low-Cost Tandem Technology: Advanced Tandem Semiconducting Polymer Cells	163
7.1.3.1	Band-Edge Energies in the Multilayer Tandem Semiconductor Polymer Structure	165
7.1.3.2	Performance of the Advanced Polymer Tandem Cell	166
7.1.4	Low-Cost Tandem Technology: Amorphous Silicon:H-Based Solar Cells	166
7.2	Organic Molecules as Solar Concentrators	169
7.3	Spectral Splitting Cells	171
7.4	Summary and Comments on Efficiency	172
7.5	A Niche Application of Concentrating Cells on Pontoons	172
8	**Third-Generation Concepts, Survey of Efficiency**	**175**
8.1	Intermediate Band Cells	175
8.2	Impact Ionization and Carrier Multiplication	177
8.2.1	Electrons and Holes in a 3D "Quantum Dot"	180
8.3	Ferromagnetic Materials for Solar Conversion	182
8.4	Efficiencies: Three Generations of Cells	185
9	**Cells for Hydrogen Generation; Aspects of Hydrogen Storage**	**187**
9.1	Intermittency of Renewable Energy	187
9.2	Electrolysis of Water	187
9.3	Efficient Photocatalytic Dissociation of Water into Hydrogen and Oxygen	188
9.3.1	Tandem Cell as Water Splitter	190
9.3.2	Possibility of a Mass Production Tandem Cell Water-Splitting Device	191
9.3.3	Possibilities for Dual-Purpose Thin-Film Tandem Cell Devices	193

9.4	The "Artificial Leaf" of Nocera	193
9.5	Hydrogen Fuel Cell Status	194
9.6	Storage and Transport of Hydrogen as a Potential Fuel	195
9.7	Surface Adsorption for Storing Hydrogen in High Density	196
9.7.1	Titanium-Decorated Carbon Nanotube Cloth	199
9.8	Economics of Hydrogen	200
9.8.1	Further Aspects of Storage and Transport of Hydrogen	200
9.8.2	Hydrogen as Potential Intermediate in U.S. Electricity Distribution	201
10	**Large-Scale Fabrication, Learning Curves, and Economics Including Storage**	**203**
10.1	Fabrication Methods Vary but Exhibit Similar Learning Curves	203
10.2	Learning Strategies for Module Cost	205
10.3	Thin-Film Cells, Nanoinks for Printing Solar Cells	207
10.4	Large-Scale Scenario Based on Thin-Film CdTe or CIGS Cells	209
10.4.1	Solar Influx, Cell Efficiency, and Size of Solar Field Required to Meet Demand	210
10.4.2	Economics of "Printing Press" CIGS or CdTe Cell Production to Satisfy U.S. Electric Demand	211
10.4.3	Projected Total Capital Need, Conditions for Profitable Private Investment	212
10.5	Comparison of Solar Power versus Wind Power	214
10.6	The Importance of Storage and Grid Management to Large-Scale Utilization	215
10.6.1	Batteries: from Lead–Acid to Lithium to Sodium Sulfur	217
10.6.2	Basics of Lithium Batteries	218
10.6.3	NiMH	220
11	**Prospects for Solar and Renewable Power**	**223**
11.1	Rapid Growth in Solar and Wind Power	223
11.2	Renewable Energy Beyond Solar and Wind	225
11.3	The Legacy World, Developing Countries, and the Third World	226
11.4	Can Energy Supply Meet Demand in the Longer Future?	227
11.4.1	The "Oil Bubble"	227
11.4.2	The "Energy Miracle"	229
	Appendix A: Exercises	**231**
	Exercises to Chapter 1	231
	Exercises to Chapter 2	232
	Exercises to Chapter 3	233
	Exercises to Chapter 4	234
	Exercises to Chapter 5	236
	Exercises to Chapter 6	236

Exercises to Chapter 7   *237*
Exercises to Chapter 8   *238*
Exercises to Chapter 9   *238*
Exercises to Chapter 10   *238*
Exercises to Chapter 11   *239*

Glossary of Abbreviations   *241*

References   *245*

Index   *251*

Preface

This book is a text on aspects of solar and renewable energy conversion based on quantum physics or "nanophysics." We take a broader view of renewable energy than is common, including deuterium-based fusion energy as approached through Tokamak-type fusion reactors. We use the physics of the sun to introduce the ideas of quantum mechanics.

Our book may be regarded as a vehicle for teaching modern and solid-state physics taking examples from the contemporary energy arena. We assume that the reader understands elementary college physics and related college-level mathematics, chemistry, and computer science. Exercises are provided for each of the 11 chapters of the book.

We omit nuclear fission power on the basis that it is available engineering, as well as that the supplies of uranium are limited.

A second view of the book is as explaining and assessing opportunities for "nanophysics" -based technology toward solving the world's looming energy problem. Earth has a population of 7 billion and rising, we are at 1 billion autos, headed toward 2 billion, with rising demand in developing nations. But oil will sharply rise in price on a scale of 30 years, the timescale on which the easily accessible oil will be used. There is definitely a problem to be solved, even without involving questions of climate change.

Fusion reactors are not usually regarded as "nanotechnology" but certainly are based on the nanophysics or quantum physics of nuclear reactions. Schrodinger's equation was used by George Gamow to explain radioactive decay, which is an inverse process to fusion. The sun would not operate without quantum mechanical tunneling of protons through Coulomb barriers. The "Tokamak" class of toroidal fusion reactors (as represented by ITER, the international fusion energy project in Cadarache, France) is the culmination of decades of fusion research with a huge accumulated literature. The complexity of this literature may have discouraged text book writers from dealing with the subject, even though the basis of the toroidal reactor is easily understood.

It is an elementary exercise in plasma physics to find that plasma containment in orbits of particles around magnetic field lines and Faraday's law of magnetic induction can lead to I^2R heating of a gas (plasma) of fusible ions having small heat capacity, at temperatures much higher than that in the sun, up to 150 million K.

A temperature of 15 million Kelvins (core of the sun) is sufficient for proton–proton fusion, powering our whole existence, only because of the high density, on the order of 150 g/cc (150 times the density of water) of hydrogen at the sun's core. This density at 15×10^6 K is unachievable terrestrially but higher temperatures are available at lower densities on the order of 10^{20} particles/m^3.

The physics of solar cells and photocatalytic production of hydrogen from water is introduced in stages: from atoms to covalent bonds to semiconductors to PN junctions. We emphasize durable thin-film solar cells that can be produced on roller-carried aluminum foil substrates in air by printing stoichiometric nanoparticles. We mention in passing that First Solar has a billion-dollar contract to build a 2 gigawatt solar cell facility in Inner Mongolia. On the other hand, we do not attempt to treat laser-based methods of terrestrial fusion, even though they may have promise.

A hindrance to interdisciplinary endeavors is the existence of compartmented literatures, such as the overwhelming literature of the Tokomak reactor, or the details of particle physics, which attest to the accumulation of knowledge but have some effect of putting walls around the knowledge. The successful worker must have the energy and audacity to plunge in to extract what is needed, overcoming barriers in names, in notation, and in choice of units, which sometimes obscure simple basic facts.

The author has benefited from teaching three classes of engineering and science graduate and undergraduate students in "Physics of Alternative Energy" at NYU Poly. In particular, he has benefited from class notes taken by Manasa Medikonda in Spring 2010. Students who have helped in this process include Angelantonio Tafuni, Karandeep Singh, Mingbo Xu, Paul-Henry Volmar, Nikita Supronova, and Diego Del'Antonio. Dell Jones of Regenesis Power is thanked for information on the lower right cover photo, of the 2 MW solar cell installation at Florida Gulf Coast University, and Dr. Karl-Heinz Haas of Fraunhofer Institute for Solar Energy is thanked for information on the upper right cover photo of a dye-sensitized flexible solar cell developed at Freiburg. The author thanks Prof. Lorcan Folan and Ms. DeShane Lyew in the Applied Physics Office for help in several ways. The assistance of Edmund Immergut, Consulting Editor, and of Vera Palmer and Ulrike Werner at Wiley-VCH, is gratefully acknowledged. Manasa Medikonda, Mahbubur Rahman, and Ankita Shah have been very helpful in preparing the manuscript. Carol Wolf, Ph.D. in mathematics and Prof. of Computer Science, has been a constant source of support in this project.

Brooklyn NY *Edward L. Wolf*
July 2012

1
A Survey of Long-Term Energy Resources

1.1
Introduction

All energy resources on earth have come from the sun, including the fossil fuel deposits that power our civilization at present. Plants grew by photosynthesis starting in the carboniferous era, about 300 million years ago, and the decay of some of these, instead of oxidizing back into the atmosphere, occurred underground in oxygen-free zones. These anaerobic decays did not release the carbon, but reduced some of the oxygen, leading to the present deposits of oil, gas, and coal. These deposits are now being depleted on a 100-year timescale, and will not be replaced. Once these accumulated deposits are depleted, no quick replenishment is possible. The energy usage will have to reduce to what will be available in the absence of the huge deposits. The words sustainable and renewable apply to this vision of the future.

There is clear evidence that the amount of available oil is limited, and is distributed only to depths of a few miles. The geology of oil very clearly indicates limited supplies. It is agreed that the continental U.S. oil supplies have mostly been depleted. Deffeyes (Deffeyes, K. (2001) "Hubbert's Peak" (Princeton Univ. Press, Princeton) authoritatively and clearly" explains that liquid oil was formed over geologic time in favored locations and only in a "window" of depths between 7500 and 15 000 feet, roughly 1.5–3 miles. (At depths more than 3 miles the temperature is too high to form liquid oil from biological residues, and natural gas forms). The limited depth and the extremely long time needed to form oil from decaying organic matter (it only occurs in particular anaerobic, oxygen-free locations, otherwise the carbon is released as gaseous carbon dioxide), support the nearly obvious conclusion that the world's accessible oil is going to run out, certainly on a timescale of 100 years.

Furthermore, scientists increasingly agree that accelerated oxidation of the coal and oil that remain, as implied by the present energy use trajectory of advanced and emerging economies, is fouling the atmosphere. Increased combustion contributes to changes in the composition of the rather slim atmosphere of the earth in a way that will alter the energy balance and raise the temperature on the earth's surface. Dramatic loss of glaciers is widely noted, in Switzerland, in the Andes Mountains, and in the polar icecaps, which relates to sea-level rises.

Nanophysics of Solar and Renewable Energy, First Edition. Edward L. Wolf.
© 2012 Wiley-VCH Verlag GmbH & Co. KGaA. Published 2012 by Wiley-VCH Verlag GmbH & Co. KGaA.

New sources of energy to replace depleting oil and gas are needed. The new energy sources will stimulate changes in related technology. An increasing premium will probably be placed on new sources and methods of use that limit emission of gases that tend to trap heat in the earth's atmosphere. New emphasis is surely to be placed on efficiency in areas of energy generation and use. Conservation and efficiency are admired goals that are being reaffirmed.

All energy comes from the sun, from the direct radiation, from the indirectly resulting winds and related hydroelectric and wave energy possibilities. These sources are considered renewable, always available. Fuels resulting from long eras of sunlight, including deposits of coal, oil, and natural gas, are nonrenewable. These resources are depleting on time scales of decades to centuries. Solar radiation is the renewable energy source that is most obviously an opportunity at present to fill the shortfall in energy.

Solar energy, while the basic source of all energy on earth, presently provides only a tiny fraction of utilized energy supply. Global energy usage (global power consumption from all sources) has been estimated as available from the solar radiation falling on 1% of the earth's desert areas. Hence, from a rational and technical point of view there need never be a lack of energy. In recent years, the oil price has been on the order of $100 per barrel, with predictions of prices much higher than the recent peak of $147 per barrel in the span of several years. From the geological point of view, the world's supply of oil is finite, and there is some consensus that in the past 100 years nearly half of it has been used. A long-term energy perspective must be based on long-term resources, and oil is not a long-term resource on a 100-year basis.

Solar energy conversion has aspects in which electronic processes are important, and for that reason this is a major topic in our book. Direct photovoltaic conversion of light photons into electron–hole pairs and into electrons traversing an external circuit is one topic of interest. The second topic, direct absorption of photons to split water into hydrogen and oxygen, will be discussed. Other permanent energy sources, which are by-products of solar energy, for instance, windpower, hydropower, and power extracted from ocean waves, do not depend in any strong way on the microscopic and nanoscopic physical processes that are the focus of our book. A key part of our book along this vein is on nuclear fusion energy, a proven resource on the sun, whose reactions are well understood. We will look carefully at several approaches to using the effectively infinite supply of deuterium in the ocean. We need technology on earth to convert the deuterium to helium as occurs on the sun, the supply of deuterium if converted to energy would supply the energy needs of our civilization for millions of years.

There are some who raise alarm at the "dangerous" suggestions that our energy-dependent civilization could be reorganized to run only on the renewable forms of energy. These observers overlap those who deny that the existing supplies of oil and coal are strictly limited, and who refuse to address the future beyond such depletions.

The strong basis for such a fear is the overwhelming dependence at present on the fossil fuels, oil, coal, and natural gas, with small amounts of hydroelectric power and nuclear power. On charts, the present consumption levels from solar power, windpower, geothermal power, wave and tidal power, are too small to be seen on the same scales.

Energy can be expressed as power times time, one kWh (kilowatt hour) is $1000 \times 3600 = 3.6 \times 10^6$ J $= 3.6 \times 10^6$ W s. The BTU, British thermal unit, is 1054 J, and the less familiar "Quad" $= 10^{15}$ BTU is thus 1.054×10^{18} J. It is stated below that the U.S. energy consumption was 94.82 Quads in 2009. In terms of average power, since a year is $365 \times 24 \times 3600$ s $= 3.15 \times 10^7$ s, this 3.17 TW. (This amounts to about 21.6% of global power, while one may note that U.S. population of 311 million is only 4.4% of the global population at 7 billion).

According to the BP Statistical Review of World Energy June 2010, the world's equivalent total power consumption in 2008 was 14.7 TW (see Figure 1.1). The largest sources in order are oil, coal, and natural gas, with hydroelectric accounting for 1.1 TW and nuclear about 0.7 TW, about 7.3 and 4.5%, respectively. Renewable power such as solar and wind are not tabulated by BP, but are clearly almost negligible on the present scale of fossil fuel power consumptions.

More details of the 2009 power consumption in the United States, breaking out the renewable energy portions, are shown in Figure 1.2.

Although the renewable energy portions are at present small, they are clearly in rapid growth. To get an idea of the growth, we find from reasonable sources

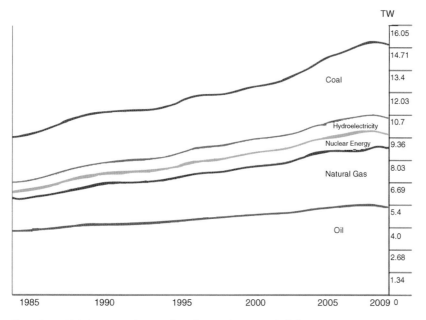

Figure 1.1 Global consumed power (based on BP Statistical Review of World Energy June 2010). The smallest band is nuclear, about 0.66 TW, and next smallest is hydroelectric, about 1.07 TW. (This is also referred to as TPES, total primary energy supply.) The largest in order are oil, coal, and natural gas, accounting for about 88.2% of all energy consumption. Astute observers agree that the three leading sources shown here are likely to significantly decrease in the next century, as prices rise due to depletion of easily available sources.

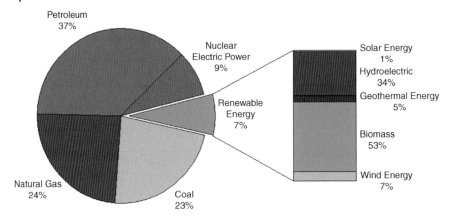

Figure 1.2 Energy consumed in United States in 2009 totals to 94.82 Quads = 9.99×10^{19} J. Of this figure, 8.16% (7.745 Quads) is classified as renewable, as broken out on the right. In the renewable category, wind accounts for 9%, thus only 0.7% of the total U.S. power consumption. (U.S. Energy Information Administration/Renewable Energy Consumption and Electricity, Preliminary Statistics, 2009).

("Renewables 2011: Global Status Report" http://www.ren21.net/Portals/97/documents/GSR/GSR2011_Master18.pdf, see also http://www.aps.org/units/gera/meetings/march10/upload/CarlsonAPS3-14-10.pdf and "Global Trends in Renewable Energy Investment 2011" (Bloomberg New Energy Finance) available at http://fs-unep-centre.org/publications/global-trends-renewable-energy-investment-2011.) estimates that in 2010 installed windpower capacity worldwide is 198 GW and growing at 30% per year. If this rate continues (which is not assured), it will be less than 20 years from 2010 until windpower reaches 5 TW, the present power from coal. This can thus be crudely extrapolated to happen by 2030. In a similar vein, in 2010 installed photovoltaic PV capacity is 40 GW and increasing at 43% per year. On this basis, it will take 13.5 years from 2010 to reach 5 TW, thus estimated in 2024.

These are long extrapolations, inherently uncertain in their accuracy. One may question that a 5 TW level from windpower is attainable from the point of view of land area and suitable sites, apart from capital investment, grid linkage and storage issues. The limiting capacities are not easy to estimate. However, one detailed study of China [1], based on windspeed data, predicted that installation of 1.5 MW turbines on mainland China could provide up to 24.7 PWh of electricity annually, which works out to an average power of 2.82 TW. This suggests that 5 TW wind capacity worldwide may be achievable. On the other hand, the *New York Times* [2] has recently published an analysis of power investment in China and finds that coal is by far the largest and most rapidly growing source of energy, and that windpower capacity is scarcely increasing.

Estimates of the power potentially available from direct photovoltaic conversion are straightforward. To reach 5 TW, assuming an average power density of 205 W/m^2 with 10% efficient solar cells requires an area $(5 \times 10^{12}/20.5)$ m^2 = 2.44×10^{11} m^2

that would be 493.8 km on a side. This area, compared to the area of the Sahara desert, 9×10^6 km^2, is 2.7%.

A detailed plan for providing renewable power to Europe has been given by Czisch. This comprehensive plan finds that transmission lines are essential to a plan that can power all of Europe at similar to present rates, without coal or oil as source (http://www.iset.uni-kassel.de/abt/w3-w/projekte/WWEC2004.pdf Dr. G. Czisch, "Low cost but totally renewable electricity supply for a huge supply area: a european/trans-european example" (http://www2.fz-juelich.de/ief/ief-ste//datapool/steforum/Czisch-Text.pdf).).

The data in Figures 1.1 and 1.2 should be regarded as accurate numbers, and this total consumption is reasonably extrapolated to double by 2050 and triple by 2100. To make a difference in the global energy pattern, any new source has to be on the scale of 1–5 TW, on a long timescale. The total geothermal power at the earth's surface is estimated as 12 TW, only a small portion extractable. It is said that total untapped hydroelectric capacity is 0.5 TW and total power from waves and tides is less than 2 TW. These latter estimates are not so certain. See "Basic Research Needs for Solar Energy Utilization," Report of the Basic Energy Sciences Workshop on Solar Energy Utilization, April 18–21, 2005, U.S. Department of Energy.

An overview of the potential renewable energy sources in the global environment has been offered by Richter. The numbers in Table 1.1 are totals and do not indicate what fractions may be extractable.

These numbers do not reflect any estimate of what portion may be extractable. Thus, Figure 1.1 indicates 1.07 TW global hydroelectric power, which is far short of 7 TW in this table for river flow energy, and elsewhere it is estimated that untapped hydroelectric power is only 0.5 TW. Such an estimate probably does not consider the potential for water turbines, analogous to wind turbines, in worldwide rivers (based on Table 8.1, Richter [3]).

Our interest is in the science and technology of long-term solutions to energy production, with emphasis on the aspects that are addressed by nanophysics, or quantum physics. Quantum physics is needed to understand the energy release in the sun and in nuclear fusion reactors such as Tokamaks on earth, and also to understand photovoltaic cells and related devices. It seems sensible to describe these

Table 1.1 Global natural power sources in terawatts (adapted from Ref. [3]).

Average global power consumed, 2008	14.7
Solar input onto land mass[a]	30 500
Wind	840
Ocean waves	56
Ocean tides	3.5
Geothermal world potential	32.2
Global photosynthesis	91
River flow energy	7

a) Solar input onto land area assuming 205 W/m^2.

processes as nanophysics, the physics that applies on the size scale of atoms and small nuclei, such as protons, deuterons, and ^{3}He. Needed also are basic aspects of materials including plasmas and semiconductors. Our hope is to provide a basic picture based on Schrodinger's equation with enough details to account for nuclear fusion reactions in plasmas and photovoltaic cells in semiconductors. From our point of view, oil, gas, coal, and nuclear fission materials are not renewable sources of energy because of the short timescales for their depletion. We focus on the energy that comes from the sun, directly as radiation, and indirectly on earth in the form of winds, waves, and hydroelectric power.

Beyond this, we consider the vast amounts of deuterium in the oceans as a sustainable source of energy, once we learn how to make fusion reactors work on earth. The heat energy in the earth, geothermal energy, is renewable but its overlap with nanophysics is not large. In a similar vein, the energy of tidal motions, which is extracted from the orbital energy of the moon around the earth, is a long-term source, but it is not strongly related to nanophysics.

The main opportunities for nanophysics are in photovoltaic cells and related devices, aspects of energy storage, and in various approaches toward fusion based on deuterium and possibly lithium. We want to learn about the nanophysical nuclear fusion energy generation in the sun for its own importance, as an existence proof for fusion, and also as a guide to how controlled fusion might be accomplished on earth.

1.1.1
Direct Solar Influx

The primary energy source for earth over billions of years has been the radiation from the sun. The properties of the sun, including its composition and energy generation mechanisms, are now known, as a result of years of research. Our purpose here is to summarize modern knowledge of the sun, with the intention of showing how the energy production of the sun requires a quantum mechanical view of the interactions of particles such as protons and neutrons at small distance scales. The Schrodinger equation, needed for understanding the rather simple tunneling processes that must occur in the sun, will be used later to get a working understanding of atoms, molecules, and solids such as semiconductors.

1.1.1.1 Properties of the Sun
The mass of the sun is $M = 1.99 \times 10^{30}$ kg, its radius $R_s = 0.696 \times 10^6$ km, at distance D_{es} about 93 million miles (1.496×10^8 km) from earth. The sun's composition by mass is approximately 73.5% hydrogen and 24.9% helium, plus a distribution of light elements up to carbon. The sun's surface temperature is 5778–5973 K, while the sun's core temperature is estimated as 15.7×10^6 K. (Much of the data for the sun have been taken from "Principles of Stellar Evolution and Nucleosynthesis" by Donald D. Clayton (University of Chicago, 1983) and "Sun Fact Sheet" by D. R. Williams (NASA, 2004)).

We are interested in the energy input to the earth by electromagnetic radiation, traveling at the speed of light, from the sun. A measurement is shown in Figure 1.3

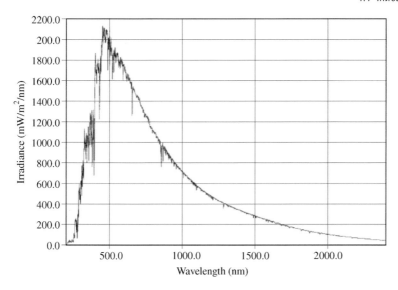

Figure 1.3 Directly measured solar energy spectrum, from 200 to 2400 nm, from a satellite-carried spectrometer just above the earth's atmosphere. The units are related to energy, mW/m² nm, and the area under this curve should be close to 1366 W/m². Note that the peak here is close to 486 nm, corresponding to a black body at 5973 K. The portion of this spectrum beyond about 700 nm cannot be seen, but represents infrared heat radiation [4].

obtained in the near vacuum above the earth's atmosphere. The curve closely fits the Planck radiation law,

$$u(\nu) = [8\pi h\nu^3/c^3][\exp(h\nu/k_B T)-1]^{-1}, \tag{1.1}$$

where $h = 6.6 \times 10^{-34}$ J s, $k_B = 1.38 \times 10^{-23}$ J/K is Boltzmann's constant, and the Kelvin temperature T is 5973 K. This is the Planck thermal *energy* density, units Joules per (Hz m³), describing the spectrum of black body radiation as a function of the frequency ν in Hertz. Equation 1.1 is the product of the number of electromagnetic modes per Hertz and per cubic meter at frequency ν, the energy per mode, and the chance that the mode is occupied. The *power* density is obtained by multiplying by $c/4$, where $c = 2.998 \times 10^8$ m/s is the speed of light. The Planck function is alternatively expressed in terms of wavelength through the relation $\nu = c/\lambda$.

Integrating this energy density over frequency and multiplying by $c/4$ leads to the Stefan–Boltzmann law for the radiation energy per unit time and per unit area from a surface at temperature T, which is

$$dU/dt = Uc/4 = \sigma_{SB} T^4, \cdots \sigma_{SB} = 2\pi^5 k_B^4/(15 h^3 c^2) = 5.67 \times 10^{-8} \text{ W/m}^2\text{K}^4. \tag{1.2}$$

The wavelength distribution of "black body radiation" peaks at wavelength λ_m such that $\lambda_m T =$ constant $= 2.9$ mm K. The value of $\lambda_m = 486$ nm for the solar spectrum is in the visible corresponding to $T \approx 5973$ K. (The sharp dips seen in Figure 1.1 attest to the wavelength resolution of the measurement, but are not central to our question of the energy input to earth. These dips are atomic absorption lines presumably from simple atoms and ions in the atmosphere surrounding the sun).

A related aspect of the radiation is the pressure it exerts, which is $U/3 = (4/3\,c)\,\sigma_{SB} T^4$. It is estimated that the temperature at the center of the sun is 1.5×10^7 K, which corresponds to radiation pressure $[4/(3 \times 3 \times 10^8)]$ s/m 5.67×10^{-8} W/m^2 K^4 $(1.5 \times 10^7 \text{ K})^4 = 0.126$ Gbar, where 1 bar = 101 kPa. This is large but a small part of the total hydrostatic pressure of 340 Gbar at the center of the sun.

The area under this curve measured above the earth's atmosphere represents 1366 W/m^2 available at all times (and over billions of years). A fraction, α (the albedo, about $\alpha = 0.3$), of this is reflected back into space. However, if we take the radius of the earth as 6371 km, then the power intercepted, neglecting α, is 1.74×10^{17} W = 174 PW (petawatts). By comparison, the worldwide power consumption, for all purposes, in 2008 was 14.7 TW, and the average total electric power usage in the United Sates in 2004 was 460 GW [5], which is only 26 parts per million (ppm) of the solar energy flux! If there are 7 billion people on the earth, this power is 24,900 kW per person. On the basis of 460 GW and 294 million persons in the United States (in 2004), the electrical power usage for 2004 was 1.56 kW per person in the United States. Worldwide total energy usage per person works out as 14.7 TW/7 billion = 2.10 kW per person.

There is thus a vast flow of energy coming from space, even after we correct for the reflected light (albedo), and the absorption effects in the atmosphere. The question of whether it can be harvested for human consumption is related to its dilute nature. At ground level in the United States, an average solar power density is about 205 W/m^2. For example, an auto at 200 HP corresponds to 200×746 watts = 14 920 W, and would require a collection area 73 m^2, much bigger than a solar panel that could be put on the roof of the car. To supply the whole country, at a conversion efficiency of 20%, a surface area of dimension about 65 miles would provide 460 GW, leaving open questions of overnight storage of energy and distribution of the energy.

The challenge is to turn the incoming solar flux (and/or other secondary sources of sun-based energy, like the wind and hydroelectric power) into usable energy on the human level. In advanced societies, it represents energy for transportation, presently indicated by the price per gallon of gasoline, and the cost per kWh of electricity.

Our second interest, in a book that focuses on nanophysics or quantum physics, that applies to objects and devices on a size scale below 100 nm or so, is to learn something about how the sun releases its energy, and to think of ways we might create a similar energy generation on earth.

The spectrum in Figure 1.3 closely resembles the shape of the Planck black body radiation spectrum, plotted versus wavelength, for 5973 K. This spectrum was measured in vacuum above the earth's atmosphere, and directly measures the huge amount of energy perpetually falling on the earth from the sun, quoted as 1366 W/m^2. If we look at the plot, with units milliwatts/(m^2 nm), the area under the curve is the power density, W/m^2. To make a rough estimate, the area is the average value, about 700 mW/(m^2 nm), times the wavelength range, about 2000 nm. So this rough estimate gives 1400 W/m^2.

This spectrum (Figure 1.3) was measured by an automated spectrometer carried in a satellite well beyond the earth's atmosphere. The sharp dips in this spectrum are atomic absorption lines, the sort of feature that can be understood only within quantum mechanics. The atoms in question are presumably in the sun's atmosphere.

We are interested in the properties of the sun that is not only the source of all renewable energy, excluding the geothermal and tidal energies and including biofuels that are grown renewably by photosynthesis, but also serves as a model for fusion reactions that might be implemented on earth. The power density at the surface of the sun can be calculated from this measured power density shown in Figure 1.3. If the radiation power density just above the earth is measured as 1366 W/m², then the power density at the *surface of the sun* can be obtained as

$$P = 1366 \, \text{W/m}^2 \times (D_{es}/R_s)^2 = 6.312 \times 10^7 \, \text{W/m}^2, \tag{1.3}$$

using the values above for the distance to the sun and the sun's radius, D_{es} and R_s, respectively. Since we have a good estimate of the sun's surface temperature T from the peak position in Figure 1.3, we can use this power density to estimate the emissivity ε, using the relation $P = \varepsilon \sigma_{SB} T^4$. This gives emissivity $\varepsilon = 0.998$, which seems reasonable.

Before we turn to an introductory discussion of how the sun stays hot, let us consider thermal radiation *from* the earth, raising the question of the energy balance for the earth itself. The earth's surface is 70% ocean, and it seems the average temperature T_E must be at least 273 K. Assuming this, the power radiated from the earth is

$$P = 4\pi R_E^2 \sigma_{SB} (T_E)^4. \tag{1.4}$$

Initially, we suppose that this power goes directly out into space. (A more accurate estimate of the earth's temperature is 288 K, see Ref. [3], p. 11.

Using $R_E = 6173$ km and taking emissivity $\varepsilon = 1$, this is $P = 160.6$ PW. Let us compare this with an estimate of the absorbed power from the sun, being more realistic by taking the Albedo (fraction reflected) as 0.3. So power absorbed is 174 PW $(1 - 0.3) = 121.8$ PW. Since the earth maintains an approximately constant temperature, this comparison indicates that a net loss discrepancy of 38.8 PW, if we neglect any heat energy coming up from the core of the earth. (It is estimated that heat flow up from the earth's center is $Q = 4.43 \times 10^{13}$ W $= 0.0443$ PW, which is relatively small. Of this, 80% is from continuing radioactive heating and 20% from "secular cooling" of the initial heat. 44.3 TW is a large number (a bit larger than shown in Table 1.1), but on the scale of the solar influx it is not important in our approximate estimate. So, we will neglect this for the moment) [6].

Thus, a straightforward estimate of power radiated from earth exceeds the well-known inflow. To resolve the discrepancy, it seems most plausible that the radiated energy does not all actually leave earth, but a portion is reflected back. A "greenhouse effect" reduces the black body radiation 160.6 PW down close to the 121.8 PW net radiation input from the sun (Figure 1.4). We can treat this as return radiation from a

Figure 1.4 Earth as seen from space, NASA. The cloud cover is evident and is a factor both in the Albedo ≈ 0.3 (the fraction of sunlight onto the earth that is reflected) and in the trapping of reradiated heat energy from the earth at 290 K (greenhouse effect). The accurate spherical shape comes from maximizing attractive gravitational energy, which caused the condensation of primordial dust into the compact, initially molten, earth. The condensation energy is estimated (see text) as $U = -0.6\, GM_E^2/R_E = -2.24 \times 10^{32}$ J, which is equal to (-1) times the present rate of global power usage times 5×10^{11} years. The power in the oceans' wave motions is estimated as 56 TW, see text. The radiation power intercepting the earth from the sun is 174 PW, which is 24.9 MW per person, on a 24 h, 7 day basis, counting 7 billion people.

"greenhouse" of temperature T_G. So the modified energy balance is

$$P = 4\pi R_E^2 \sigma_{SB}[(T_E)^4 - (T_G)^4] = 121.8\ \text{PW}, \tag{1.5}$$

where we have taken the "greenhouse" temperature T_G as 191.3 K, in a simple analysis. According to Richter (op. cit., p. 13), the most important greenhouse gases are CO_2 and water vapor [3].

1.1.1.2 An Introduction to Fusion Reactions on the Sun

In the simplest terms, the power density $P = 63$ MW/m^2 leaving the surface of the sun comes from nuclear fusion of protons, to create ^{4}He, in the core of the sun. Let us find the total power radiated by the sun. This is $4\pi R_s^2 \times 63.12$ MW $= 3.82 \times 10^{26}$ W, making use of $R_s = 0.696 \times 10^6$ km. This 3.82×10^{26} W is such a large value, do we need fear the sun will soon be depleted? Fortunately, we can be reassured that the lifetime of the sun is still going to be long, by estimating its loss of mass from the radiated energy. Using the energy–mass equivalence of Einstein,

$$\Delta M c^2 = \Delta E, \tag{1.6}$$

on a yearly basis, we have $\Delta E = 3.82 \times 10^{26}$ W $\times$ 3.15×10^7 s/year $= 1.20 \times 10^{34}$ J/year. This is equivalent to $\Delta M = (1.20 \times 10^{34}$ J/year$)/c^2 = 1.337 \times 10^{17}$ kg/year. Although ΔM is large, it is tiny in comparison to the much larger mass of the sun, $M = 1.99 \times 10^{30}$ kg. Thus, we find that the fractional loss of mass per year, $\Delta M/M$, for the sun is 1.337×10^{17} kg/year $\div$ 1.99×10^{30} kg $= 6.72 \times 10^{-14}$/year. This is tiny indeed, so the radiation is not seriously depleting the sun's mass. On a scale of 5.4 billion years, the accepted age of the earth, the fractional loss of mass of the sun, during the whole lifetime of earth, taking the simplest approach, has been only 0.036%.

Where does all this energy come from? It originates in the "strong force" of nucleons, which is large but of short range, a few femtometers. Chemical reactions deal with the covalent bonding force, nuclear reactions originate in the strong force, about a million times larger. The energy is from burning hydrogen to make helium, in principle similar to burning hydrogen to make water, but the energy scale is a million times larger.

In more detail, the composition of the sun is stated as 73.5% H and 24.9% He by mass, so the obvious candidate fusion reaction is the conversion of H into He. The basic proton–proton fusion cycle leading to helium in the *core* of the sun (out to about 0.25 of its radius) has several steps that can be summarized as

$$4p \rightarrow {}^4\text{He} + 2e^+ + 2\nu_e. \tag{1.7}$$

This says that four protons lead finally to an alpha particle (two protons and two neutrons, which forms the nucleus of the Helium atom), two positive electrons, and two neutrino particles.

This is a fusion reaction of some of the elementary particles of nature, which include, besides protons and neutrons, positive electrons (positrons) and neutrinos ν_e. Positrons and neutrinos may be unfamiliar, but a danger is to become intimidated by unnecessary details, rather than, in an interdisciplinary field, to learn and make use of essential aspects. The important aspect here is that energy is released when particles combine to form products the sum of whose masses are less than the masses of the constituents. Furthermore, as we will learn, this reaction can proceed only when the source particles have high kinetic energy, to overcome Coulomb repulsion when the charged particles coalesce. In addition, the essential process of "quantum mechanical tunneling," an aspect of the wave nature of matter, allows the reaction to proceed when the interparticle energies are in the kiloelectron volt (keV) range, available at temperatures above 15 million K. From elementary physics, we recall that the average kinetic energy per degree of freedom in equilibrium at temperature T is

$$E_{av} = \tfrac{1}{2} k_B T, \tag{1.8}$$

where Boltzmann's constant $k_B = 1.38 \times 10^{-23}$ J/K. The energy units for atomic processes are conveniently expressed as electron volts, such that 1 eV $= 1.6 \times 10^{-19}$

$J = 1.6 \times 10^{-19}$ W s. Chemical reactions release energy on the order of 1 eV per atom, while nuclear reactions release energies on the order of 1 MeV per atom, see Figure 1.5. A broad distribution of particle speed v is allowed in the normalized Maxwell–Boltzmann speed distribution,

$$D(v) = (m/2\pi k_B T)^{3/2} 4\pi v^2 \exp(-mv^2/2 k_B T). \tag{1.9}$$

While one may have learned of this in connection with the speeds of oxygen molecules in air, it usefully applies to the motions of protons at 15 million K in the core of the sun.

The most probable speed is $(2\,kT/m)^{1/2}$ that corresponds to a kinetic energy $E_k = 1/2\,mv^2$ of kT. In connection with the probability of tunneling through the Coulomb barrier, which rises rapidly with rising interparticle energy (particle speed), one sees that the high-speed tail of the Maxwell–Boltzmann speed distribution is important. The overlap of the speed distribution, falling with energy, and the tunneling probability, rising with energy, typically as $\exp[-(E_G/E_k)^{1/2}]$ as we will learn later, leads to what is known as the "Gamow peak" for fusion reactions in the sun. (The sun's neutrino output has been measured on earth, and is now regarded as in satisfactory agreement with the p–p reaction rate in the core of the sun [9].)

The energy release of this reaction can be calculated from the change in the $m_i c^2$ terms. Using atomic mass units u, we go from 4×1.0078 to $4.0026 + 2\,(1/1836) = 9.51 \times 10^{-3}$ u, and using 935.1 MeV as uc^2, we find 8.89 MeV per ^{4}He, neglecting the neutrino energy. The atomic mass unit u is nearly the proton mass, but defined in fact as 1/12 the mass of the carbon 12 nucleus.

We should point out the large scale of the fusion energy release, here nearly 9 MeV on a single atom basis. This is about a million times larger than a typical chemical reaction, on a single molecule basis. The nuclear force that binds the protons and neutrons in the nuclei is indeed about a million times stronger than the typical

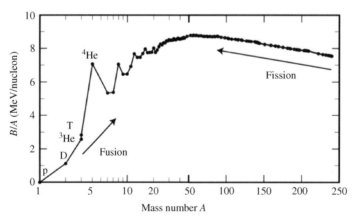

Figure 1.5 The sun's radiating power comes largely from nuclear fusion of protons p into ^{4}He at 15 million K. Mass (nucleon) number $A = Z + N$. p, D, and T, are equivalent, respectively, to ^{1}H, ^{2}H, and ^{3}H. (reproduced from Ref. [8], Figure 1).

covalent bond energies in molecules and solids. This large size is, of course, a driving factor toward the use of fusion reactors on earth.

Returning to the sun, it is believed that the p–p cycle accounts for about 98% of the sun's energy output [10], all occurring in the core. The energy diffuses slowly out to the outer surface with attendant reductions in pressure and temperature, the latter from 15 million K to about 5800 K.

The first reaction in the proton–proton cycle at the sun's core is [11]

$$p + p \rightarrow D + e^+ + \nu_e, \tag{1.10}$$

where D is the deuteron, the bound state of the neutron and proton, which has mass 2.0136 u. (The mass unit, u, is defined as 1/12 of the mass of the ^{12}C nucleus. One u is about 1.67×10^{-27} kg). Here, the energy release is 1.44 MeV, which includes 0.27 MeV to the neutrino.

This first proton–proton reaction occurs very frequently in the sun, as the first step in the basic energy release process. But this reaction is impossible from the point of view of classical physics. It should not occur, from the following reasoning. Accepting the estimated temperature at the center of the sun as 1.5×10^7 K, the thermal energy in the center of mass motion of two protons would be $1/2\, k_B T = 1/2\, 1.38 \times 10^{-23} \times 1.5 \times 10^7$ J $= 1.035 \times 10^{-16}$ J $= 646.9$ eV.

(There will more realistically be a distribution of kinetic energies, and energies higher than 10 keV will frequently be available to colliding protons at 15 million K). But any such estimated energy is far short of the potential energy $k_C e^2/r$ that is required classically to put two protons in contact. (Here $k_C = 9 \times 10^9$ and $e = 1.6 \times 10^{-19}$ C). The radius of the proton has been measured and we will take it as 1.2×10^{-15} m. In this case, the Coulomb energy $k_C e^2/r$ in eV is $9 \times 10^9 \times 1.6 \times 10^{-19}/(2 \times 1.2 \times 10^{-15}) = 0.6$ MeV. This energy is vastly higher than the kinetic energy (see Figure 1.6). Classically, this reaction will not occur because the two protons will never come into contact.

This fundamental discrepancy was resolved in the early years of the quantum mechanics, and in particular by George Gamow [12], an American physicist. The resolution is that the reaction proceeds by a process of "quantum mechanical tunneling," and the kinetic energies near the "solar Gamow peak" in the range 15–27 keV provide most of the reactions. We will return to this topic later. The process is now completely understood, and we will explore it in some detail because it is also central to experimental approaches to generating fusion energy on earth.

A later and important reaction in the p–p cycle, which we will come back to, is fusion of two deuterons. The result can be a triton T plus a proton, ^{3}He plus a neutron, or an α (^{4}He) plus a gamma ray (photon). (A triton is one proton plus two neutrons, and forms tritium atoms similar to hydrogen and deuterium atoms. Tritium, as opposed to deuterium, does not occur in nature).

1.1.1.3 Distribution of Solar Influx for Conversion

The sun's energy density varies considerably with differing cloud cover characteristic of different parts of the world. A summary of this is shown in Figure 1.7a. The

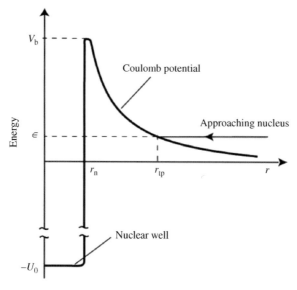

Figure 1.6 Sketch of fusion by tunneling through Coulomb barrier. Even at 15 million K, the interparticle energy ε (kilovolts) is far below the Coulomb barrier V_B (megavolts). (reproduced from Ref. [8], Figure 2).

squares marked on this map represent about 0.16% of the earth area and are judged to be sources of all the world's power need, about 20 TW estimated for mid-century (see Figure 1.1), assuming the areas are covered with 10% efficient solar cells [8]. This is total power consumed, not just electric power!

The units in Figure 1.7b are effective hours of sunlight per year on a flat plate collector, including weather effects. The peak value 2100 h per year of sunlight works out an average W/m² value as $2100/(365 \times 24)$ 1000 W/m² = 240 W/m². Note that in the Midwest portion of the United States where the effective hours per year are shown as around 1600, this corresponds to $1600/365 = 4.4$ h per day, at around 1000 W/m². This time span, 4.4 h, is roughly the duration of the peak electric demand, often about twice the night-time demand.

1.1.2
Secondary Solar-Driven Sources

Wind energy and river flow energy are indirect results of heating by the sun. A map of wind speed in the United States is shown in Figure 1.8. The peak values are in the range 8–9 m/s. The uneven distribution of the resource makes clear the need for a wide grid network or for conversion to a fuel such as hydrogen that could be piped or shipped in containers.

1.1.2.1 Flow Energy
The power that can be derived from wind or water flow is proportional to v^3. To understand this result, consider an area $A = \pi R^2$ oriented perpendicular to a flow at

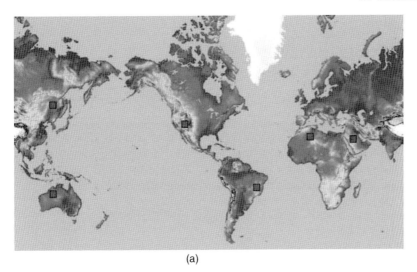

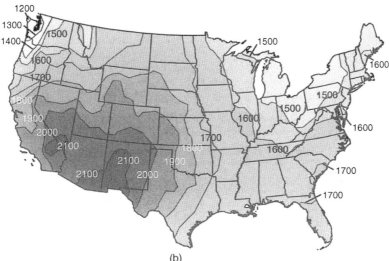

Figure 1.7 (a) Map of requirement of land to meet the world's total power demand (http://ethic-forum.unife.it/E602373e_ev-Balzani.pdf.) in mid-century solely by solar cells. (b) Map of solar intensity across the United States. The units are described in the text. (Reproduced from Ref. [3], Figure 13.5, p. 159. Source U.S. Department of Energy, Energy Efficiency and Renewable Energy Division).

speed v of fluid of density ϱ. In one second, a length $L = v$ containing mass $M = Av\varrho$ will pass through the area. This represents a flow of kinetic energy $dK/dt = dM/dt\, v^2/2$, so that power $P = \eta\, dK/dt = \eta\, dM/dt\, v^2/2$ can be obtained if the efficiency of the turbine is η. Thus,

$$P(R) = \eta \pi R^2 \varrho v^3 / 2. \tag{1.11}$$

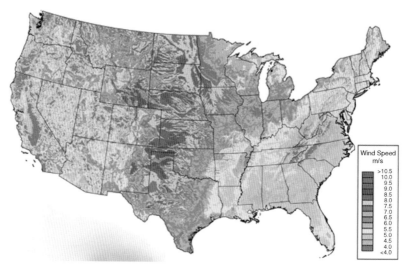

Figure 1.8 Distribution of wind speeds across the United States (U.S. Department of Energy).

A turbine such as the one shown in Figure 1.9 with radius $R = 63.5$ m, and assuming $v = 8$ m/s, taking $\varrho = 1.2$ kg/m^3 for air at 20 °C yields

$$P = \eta \pi \, 63.5^2 \, 1.2 \, 8^3 / 2 = \eta \, 3.89 \text{ MW}.$$

The best efficiency in practice is about 0.4, giving 1.56 MW/turbine at the assumed 8 m/s, which is a favorable value, shown in the dark areas of the wind map, Figure 1.8, typically in the United States in a band running from Texas to Minnesota.

It is quite easy to show that the maximum efficiency is about 0.59 (Betz's law) (http://c21.phas.ubc.ca/article/wind-turbines-betz-law-explained.) by realizing that the speed v' behind the turbine is reduced, and the average speed is $v_{av} = \frac{1}{2}(v + v')$. Thus, the corrected formula is

$$P(R) = \pi R^2 \varrho v_{av}(v^2 - v'^2)/2. \tag{1.12}$$

This formula provides a maximum power at most 0.59 of the unperturbed power $P_0(R) = \pi R^2 \varrho v^3 / 2$. This corresponds to $v = v/3$, so one can see why the wind turbines are not longitudinally arranged because the exit air velocity is quite reduced.

Consider an array of such turbines, spaced by 10 R. Then the power per unit ground area delivered by the array of the designated turbines at 8 m/s is 1.56 MW/$(635 \text{ m})^2 = 3.86$ W/m^2. A rough comparison with solar cells is that an average solar power at earth is 205 W/m^2 with an expected efficiency around 0.15, thus 30.75 W/m^2. The possibility exists of having both solar cells and wind turbines in the same area, plausible if the area is not cultivated. Questions of the installation costs are deferred, but the starting estimate of $1/(peak installed watt) generally is useful.

We can ask how large a windfarm is needed to generate 500 GW, approximately the electricity used in the United States? If we take 3.86 W/m^2, the answer is

Figure 1.9 Enercon model E-126 7.5 MW wind turbine. The hub height is 135 m. The specifications say that the machine can be set to cut off at a chosen wind speed in the range 28–34 m/s. From the text one would extrapolate to a power from one device, at 28 m/s, of 66.9 MW. The specifications say the blades are epoxy resin with integrated lightning protection (http://www.enercon.de/en-en/66.htm.).

Area = $500 \times 10^9/3.86 = 12.95 \times 10^{10}$ m², or 360 km or 223 miles, on a side. This is comparable to the area of the state of Iowa, which is equivalent to 237 miles on a side! The positive aspect is that the turbines do not necessarily preclude the normal use of the land, for example, to grow wheat or corn. But there is no escape from the reality that both wind energy and solar energy are diffuse sources.

Or we may ask how many wind turbines at 1.56 MW per turbine? That number is $N = 500 \times 10^9/1.56 \times 10^6 = 320\,513$ turbines. At a spacing of 635 m = 0.394 miles per turbine or 2.53 turbines per mile, we could imagine turbines along 126 684 miles of highway. The total mileage in the Interstate Highway system is 46 751 miles, while the total U.S. highways extend 162 156 miles. The cost of the turbines at $1 per Watt is $500 billion. The cost of the U.S. Interstate Highway system is said to be $425 billion in 2006 dollars (http://en.wikipedia.org/wiki/Interstate_Highway_System.). $500 billion is approximately equal to 0.07 of the U.S. military budget for a period of 10 years.

The same kinetic energy extraction analysis applies to water flow in a river, which benefits immediately from the factor 1000/1.2 = 833 increase in density. A recent measurement of Mississippi water flow (http://blog.gulflive.com/mississippi-press-news/2011/05/mississippi_river_flooding_vic.html.) recorded 11 mph velocity and 16 million gallons per second flow under a bridge near Vicksburg, MS. With conversions 1 mph = 0.447 m/s and 1 U.S. gallon = 4.404×10^{-3} m³, we have 4.92 m/s and $dV/dt = 7.06 \times 10^4$ m³/s for water flow at this location. The power is then $(dM/dt)(v^2/2) = 1000$ kg/m³ $(7.06 \times 10^4$ m³/s$)(4.92$ m/s$)^2/2 = 94.8$ MW.

The efficiency can be at most 0.59, corresponding to loss of speed by 2/3, and the resulting disruption of the river flow if the full cross section were filled with rotor blades would be prohibitive. Still it seems that tens of MW could be extracted from such a flow if it were continuous and if the installations could be sited to avoid blocking of commerce.

The size of a 1 MW river-flow or tidal-flow turbine is much smaller than a 1 MW wind turbine because of the 1000-fold increase in water density. Probably, this means the water turbine would be cheaper. Water turbines, highly developed for hydroelectric installations, in smaller forms for river-flow applications are not as well established commercially as are wind turbines.

1.1.2.2 Hydroelectric Power

Water running through turbines is used to generate electricity, with a typical efficiency of 90%. It is evident from Figure 1.2 that hydroelectric power is at present by far the largest renewable energy source, amounting to about 1.07 TW worldwide in 2008, or about 7.3%. These are extremely large projects typically, and the easiest sites are already utilized (see Figure 1.10). The situation, often, for a large installation is that it is close to a copper mine or an aluminum smelting facility, which has supported the capital investment. The availability of efficient DC power transmission lines may make the benefit of these large installations more widely available.

Similar large facilities are at Niagara Falls in the United States and the Akosombo Dam in Ghana, Africa. The Three Gorges dam in China at completion has a capacity of 22.5 GW. The planned Grand Inga Dam in Congo is projected as 39 GW. The Belo

Figure 1.10 Grand Coulee Dam is a hydroelectric gravity dam on the Columbia River in the U.S. state of Washington. The dam supplies four power stations with an installed capacity of 6.81 GW. It is the largest electric power-producing facility in the United States (http://en.wikipedia.org/wiki/File: Grand_Coulee_Dam.jpg.).

Monte Dam on the Xingu, a tributary of the Amazon, has been approved (http://www.bbc.co.uk/news/world-latin-america-13614684.) by Brazil. The dam would be 3.7 mi long and the power would be 11 GW. The Itaipu Dam between Brazil and Paraguay is rated at 14 GW. From 20 0.7 GW generators, two 600 kV HVDC lines, each about 800 km long, carry the DC power to Sao Paolo, where terminal equipment converts to 60 Hz. It provides 90% of electric power in Paraguay and 19% of power in Brazil (http://en.wikipedia.org/wiki/Itaipu_Dam.).

Turbines can be made with the capacity to be reversed and to pump water back to the reservoir when demand is low. This storage capability is called "pumped hydro" and efficiency in the pumping mode can be 80%. Capacity on the American and Canadian sides of the Niagara River totals 5.03 GW, of which 0.374 GW is pumped storage/power producing units (pumped hydro) such as shown in the next figure. The pumped storage facility Carters Dam in Georgia provides a maximum power output of 500 MW during peak demand conditions. Figure 1.11 shows the generators and power distribution from this large water reservoir created by an earthen dam (http://en.wikipedia.org/wiki/File:U.S.ACE_Carters_Dam_powerhoU.S.e.jpg (http://www.niagarafrontier.com/power.html).).

Figure 1.11 Illustration of a 500 MW pumped hydroelectric energy storage facility. The turbines can be reversed to pump water back into the reservoir. This form of energy storage in the electric grid is of larger capacity and lower cost than any known form of battery. Carters Dam in Georgia, U.S. (http://en.wikipedia.org/wiki/File:U.S.ACE_Carters_Dam_powerhoU.S.e.jpg (http://www.niagarafrontier.com/power.html).).

1.1.2.3 Ocean Waves

According to Table 1.1, the power available in all the oceans' waves is 56 TW, about 3.8 times the global energy consumption at present. Since the area of ocean is 139.4×10^6 mi^2 = 3.61×10^{14} m^2, mi = 1609 m, the power per unit area from this estimate is 0.155 W/m^2. This seems small, but of course ocean waves are really a secondary result of winds, which are themselves a secondary result of the sun's heating.

To check such an arbitrary number, a scientist or technologist should be skeptical and might seek to test it against his own rough estimate.

Estimate of Wave Energy Suppose the average ocean wave amplitude D is 1 m and the average frequency $\omega = 2\pi f$ of the oscillation is 0.1 rad/s. So a wave passes a given location every $1/f = 62.8$ s. (These are guesses on the average depth and frequency of ocean waves). In the simplest model of an ocean wave, the water moves vertically in simple harmonic motion, $y = D \sin \omega t$. The speed dy/dt is thus $-D\omega \cos \omega t$, with maximum speed $D\omega$. The energy of this oscillation is, thinking of M as the mass of all the ocean to a depth 1 m,

$$E = 1/2 M (D\omega)^2 \tag{1.13}$$

so that the power dE/dt is

$$P = 1/2 M (D\omega)^2 \omega. \tag{1.14}$$

Thus, $P = 1/2\, M\, (1 \times 0.1)^2\, 0.1$, where $M = A\, D \times 1000$ kg, where A is the ocean area and 1000 is the density of water in kg/m³. Thus,

$$P/A = 0.5 \times 1000 \times 10^{-3} = 0.5\ \text{W/m}^2,$$

compared to 0.155 W/m² from Table 1.1. We predict the total power in ocean waves is 181 TW, on this crude estimate, compared to 56 TW from Table 1.1.

This crude estimate is closer than one might have expected! (It is likely that the typical frequency is higher, and the typical amplitude is smaller). The estimate also helps us understand that the power is proportional to the square of the wave height and the cube of the wave frequency. In fact, the trajectory of water particles as the wave passes is not vertical but circular, and the wave is mathematically a "trochoidal" wave rather than a sinusoidal wave. If one imagines a disk of radius R rolling, a point on the radius $r = R$ experiences sinusoidal motion but a point at radius $r < R$ executes trochoidal motion. The model of sinusoidal motion is still useful (http://hyperphysics.phy-astr.gsu.edu/hbase/waves/watwav2.html.).

The designs of devices, termed wave energy converters, WEC, to extract the wave energy, are naturally adapted to a particular situation, such as at a given depth of water beyond a shoreline, where waves are approaching land. The wave amplitude and speed increase as the open water wave approaches land. Water depths in the range 40–100 m are typical of present installations [13].

The potential extractable wave energy from the Pacific west coast of the United States is estimated [13] as 255 TWh per year, and in Europe about 280 TWh per year. These numbers are equivalent to powers of 0.029 TW and 0.032 TW, respectively (29 GW is an appreciable fraction, about 0.06, of electric power consumption in the United States). It is not clear what the capital and operational costs of such extraction would be, but at least one commercial device, the Pelamis, has been subsidized by the government of Portugal and put into service.

A plausible estimate of available wave power along a coastline is in terms of power per unit length of the coastline. On the Atlantic coast of Great Britain this is estimated [14] as 40 kW/m of exposed coastline. This estimate depends on the height of the waves, which is a function of the windspeed and the unimpeded span of water facing the coast over which the waves can collect energy from the wind. This estimate might be compared to the estimate above for the Pacific coast of the United States. If that coastline is 1000 miles or 1.6 Mm, then we get, at 40 kW/m, the estimate 64 GW, fairly close in agreement.

As waves approach land at depth d, the wave speed is

$$V = [(g\lambda/2\pi)\tan h(2\pi d/\lambda)]^{1/2}. \tag{1.15}$$

The Pelamis (the word means "water snake") device is a linear array of four linked pontoons, each 30 m long, oriented perpendicular to the waves. The flexing motion occurring at the linking joints with wave passage is used to create electricity. Pelamis devices totaling 2.25 MW capacity have been installed in the sea near Portugal. Vertically bobbing buoy devices anchored at modest depths are also practical. Devices may also be based on trapping water from the tops of waves, extracting energy as that water falls back into the sea.

While the potential seems appreciable for tapping wave energy in coastal regions, the much larger potential power at the open sea seems in practice unavailable, by virtue of its remoteness.

On the other hand, one might conceive of "ocean stations," large floating facilities, which need not be close to land. Such stations might be used, for example, as a basis for desalinization of seawater, for extraction of deuterium from the sea, or for electrolytic hydrogen generation. Possible "ocean stations" for hydrogen production could also harvest wind and solar power. Schemes for delivery by tanker, analogous to the shipping of oil and liquid natural gas, might evolve.

An "ocean station" seems more practical than a "space station" for the human future, let alone facilities discussed (in the United States) for colonization of the moon. We will return in Chapter 5 to an estimate of the cost of a satellite system to send solar energy to earth from space.

An economically sound and competent city might launch its own ocean station, to capture energy for its sphere of influence, and thus reduce dependence on its surrounding grid. This scenario might extend to viable coastal cities worldwide, perhaps Dhaka or Mumbai, beyond New York City.

1.1.3
Earth-Based Long-Term Energy Resources

Some of the long-term energy that is available is stored in the earth, or is the result of the orbital motion of the moon around the earth. In addition, the composition of the ocean contains enough deuterium, present from the beginning of the earth, to constitute a long-term resource.

1.1.3.1 Lunar Ocean Tidal Motion
Tides are caused by the motion of the moon around the earth, in large part. In "funnel" locations like the Bay of Fundy, the flows can be large and rapid. Harvesting tidal flows can be similar to harvesting the flow energy of a river. In some cases, all of the flow can be funneled into a single set of turbines, a situation more like that at Niagara Falls. This is suggested by the artificial tidepool shown here in Figure 1.12.

Famous optimum locations, such as the Bay of Fundy, which has a tidal range of 17 m, are at least partly exploited. At present, the 20 MW tidal power plant at the Bay of Fundy is the only such plant in operation. However, there is scope for more energy to be tapped in this category.

An example of a much larger potential is shown in Figure 1.13, based on tidal flows in the British Isles. In the analogy to the tidal basin, the North Sea roughly plays that role in the example of the British Isles as the gateway between the Atlantic and the North Sea. The energy flow can be taxed on the intake and the exhaust of the cycle.

Calculations of the available power, up to 190 GW, are indicated on the diagram.

On smaller scales, the common occurrence of sandbars parallel to a beach suggests many locations that could be utilized. In the Atlantic coast of the United States, the Outer Banks of North Carolina enclose Pamlico Sound, an area about 2000 square miles, or $5.18 \times 10^9 \, m^2$. The tidal excursion at Cape Hatteras, on the ocean side, is

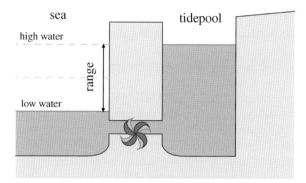

Figure 1.12 An artificial tide pool [15]. As shown the pool is filled by the high tide at an earlier time, and is now able to discharge water through a turbine generating electricity.

3.6 feet, while the tidal excursion on the inside, for example, at Rodanth, on Pamlico Sound is only 0.72 feet. So it appears that the interior, Pamlico Sound, is decoupled from the tidal excursion on the Atlantic side, by the relatively small openings, through the Outer Banks, between the Sound and the open Atlantic Ocean, where the tides are over 3 feet. Nonetheless, the energy exchanged every 12 h, $U = Mgh$, where M is the mass of the water in Pamlico Sound to a depth of $h = 0.72$ feet, we can estimate to be

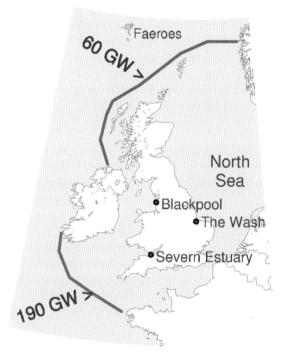

Figure 1.13 Map suggesting locations of optimal tidal energy flows from the Atlantic Coasts of the British Isles [16].

quite large. Namely, $U = (5.18 \times 10^9 \, \text{m}^2 \times 0.22 \, \text{m}) \times 1000 \times 9.8 \times 0.22 = 2.46 \times 10^{12}$ J. In terms of an average power $P = dU/dt$, this is 57 MW. The annual market value of the entirety of this potential power, at \$0.14/kWh, would be $57 \times 10^6 \times 3.15 \times 10^7 \times (3.6 \times 10^6)^{-1} \times 0.14 = \$69.8 \times 10^6$. In an age of governments needing to raise taxes, this might be an incentive to install water turbines.

This situation is present in numerous smaller scale examples. In New York City, the TV host will speak of the danger, on a given day, at a particular beach, of "rip currents," to swimmers. "Rip currents" are tidal flows of water through such constrictions (between open sea and a tidal pool) as we have discussed. There are many locations where sandbars or "keys" are located just off the mainland.

It would seem that constructing artificial entrapments of this sort, for example, a sandbar ("key") extended by levees (dams) to trap tidal flows, augmented with water turbines and grid connections, could be a new activity for the illustrious U.S. Army Corp of Engineers, which has installed numerous bridges, levees, and other water-related engineering projects in the United States.

1.1.3.2 Geothermal Energy

Geothermal potential according to Table 1.1 is 32.2 TW with a higher value, 44.3 TW, from a different source [17]. The core of the earth is molten, and heat leaks out to the surface. The energy release actually comes from two sources. One is radioactive decay of elements like uranium and thorium in the outer layers of the earth. The second is the heat from the earth's core that remains molten, at a much higher temperature. While the trend is of cooling of the core from its primordial high temperature, it has been mentioned that some heat input, a continual heating of the earth's core, comes from the motion of the moon, which continually distorts the shape of the earth, as well as driving the tides.

From a physics point of view, the condensation energy in forming the earth from a dispersed cloud of dust to a condensed sphere of radius R,

$$E = -3/5 \, GM^2/R, \tag{1.16}$$

is a benchmark value, easily calculated. Here G is the universal gravitation constant, $G = 6.67 \times 10^{-11}$, so that, with $M = 5.97 \times 10^{24}$ kg, and $R = 6.37 \times 10^6$ m, we find

$$E = -2.24 \times 10^{32} \text{ J}.$$

This energy, released as kinetic energy, is vast, comparable to the present rate of consumption extended for 4.84×10^{11} years! It is clear that most of this energy has already been lost, mostly by radiation shortly after the condensation. If we were to attribute this full energy to heating of the earth, we can estimate what the temperature would have been. In a simple model of a solid or liquid, the thermal energy is

$$U = 3Nk_B T. \tag{1.17}$$

If we attribute all the mass $M = 5.97 \times 10^{24}$ kg to iron atoms, atomic mass $55.85 \times 1.67 \times 10^{-27}$ kg, then $N = 6.4 \times 10^{49}$ atoms, and $T = U/3Nk_B = 84.5 \times 10^3$ K. The radiation power from the surface of the early earth at that temperature would be

$P = 4\pi R^2 \sigma_{SB} T^4$, where the Stefan–Boltzmann constant $\sigma_{SB} = 5.67 \times 10^{-8}\, \text{W/m}^2\, \text{K}^4$. This is evaluated as

$$P = 1.47 \times 10^{27}\, \text{W}.$$

In the simplest view, this suggests that the original heat energy could be radiated away in about 42 h, since $42 \times 3600 \times P = U$. But the radiative cooling quickly slows as the temperature falls, and the linear approach fails. At present, the inner core temperature has been estimated as 5700 K, while lava (magma) at temperatures ~1500 K is present at some locations as close as 10 km to the earth's surface. The remaining heat energy in the earth's core is, of course, enormous and certainly can be regarded as a renewable resource.

Practical extraction of the earth's heat is accomplished at locations where molten lava extends close to the surface, providing regions of hot rock that are used to heat injected water to produce steam. U.S. capacity of this type is 3.09 GW, with the largest facility at The Geysers field (http://www.gwpc.org/meetings/forum/2007/proceedings/Papers/Khan,%20Ali%20Paper.pdf.) in CA. Iceland has exploited its geothermal energy to a great extent. A map (http://www.magma-power.com/pages/magma_power_plant.html.) of locations in the United States where magma exists within 10 km of the surface reveals sites concentrated in western states and along the Aleutian Islands. While plant designs have been offered for tapping directly into a lava field, this has not been accomplished.

1.1.3.3 The Earth's Deuterium and its Potential

Fusion of light elements to release energy is the heating mechanism of the sun. A good starting point for fusion is the deuteron, two of which can fuse to make ^{4}He with release of nearly 24 MeV of energy. The most likely products for DD fusion are actually a triton plus a proton, with 4 MeV; or ^{3}He plus a neutron, with 3.27 MeV, so that the average energy release per DD fusion is 3.7 MeV. The deuteron fusion reactions are considered important because D particles, Deuterons, are present on earth, notably in seawater. Wherever protons occur, there is about 1/6400 chance of finding instead a Deuteron. Heavy water HDO, therefore, occurs as $1/3200 = 0.031\%$ of all water. There is enough in the ocean that this is considered a sustainable or renewable energy source. The problem is that at present there is no practical process using Deuterons to actually release energy by the fusion reactions.

If we take the ocean mass as 1.37×10^{21} kg, comparing it with the mass per water molecule, $18 \times 1.67 \times 10^{-27}$ kg, we find that there are $N = 4.6 \times 10^{46}$ water molecules in the ocean. This means there are 9.2×10^{46} H atoms, and therefore there are 1.42×10^{43} deuterons. The energy release, if all of these deuterons were fused at the average energy release of 3.7 MeV, is therefore $1.42 \times 10^{43} \times 3.27 \times 10^6 \times 1.6 \times 10^{-19}\, \text{J} = 7.45 \times 10^{30}\, \text{J}$. If the present energy consumption is 14.7 TW, so that one year's energy consumption is 4.63×10^{20} J, the deuteron-based energy would last for 1.6×10^{10} years. So, we may say that the deuterium in the ocean, if it can be converted, is a renewable resource. In Chapter 4 of the book, we will look into the possibilities for achieving this release of energy.

1.1.4
Plan of This Book

The underlying physics of solar energy, with a fairly detailed account of how the sun delivers its energy to earth are treated in Chapter 2. To prepare the reader for nanophysics-based energy conversion devices, principally solar cells of various types, background is provided in Chapter 3. Chapter 4 explains three methods that are known to release fusion energy in laboratory situations on earth. The power output from a Tokamak-type fusion reactor is analyzed and numerically estimated by scaling the simplified reaction model, shown in Chapter 2, to predict the sun's output, to the Tokamak realm of parameters. The topics then turn, in Chapter 5, to exploiting the solar radiation input to earth, converting some of the energy to electricity. The physics of solar thermal energy conversion is compared to that of photovoltaic conversion, and a survey of solar cell types is presented. Chapters 6–8 deal in more detail with types of solar cells, including prospects for developing new cells with higher efficiency and possibly at lower cost. Chapter 9 deals with aspects of producing hydrogen gas by photocatalytic cells, as well as practical possibilities for making hydrogen a storage medium for energy produced by wind or solar power. Chapter 10 deals with manufacturing and economic aspects of solar power, with attention to processes that might be scalable to large volume and low cost to replace a significant fraction of the power now obtained from oil, natural gas, and coal. Finally, Chapter 11 deals with the future of renewable energy, as a part of the global energy future.

2
Physics of Nuclear Fusion: the Source of all Solar-Related Energy

Energy from the sun is entirely produced by nuclear fusion, so it is important to understand how this works. There are two main aspects of a complex process that are basic. First is the strong nuclear attractive force leading to reactions of the elementary proton particles and to formation of more stable nuclear products, such as ^{4}He (see Figure 1.5). Since this combination of the four nucleons (two protons and two neutrons) is more stable, a large kinetic energy Q is released. The strong interaction binding in ^{4}He amounts to around 9 MeV per particle. The strong nuclear force has a short range, on the order of a femtometer $f = 10^{-15}$ m. This force acts equally binding protons and neutrons independent of the electric charge. The individual particles have "charge radii" on the order of femtometers. The charge radius is the size within which the electric charge is contained, about 1.2 f for a proton, directly measured by electron scattering at high energy. (These experiments actually show substructures of the proton and neutron, each built up from three smaller particles, called quarks, within the overall radius 1.2 f. These details are not needed for our discussion.) Nuclei in general have a radius, measured by high-energy scattering experiments, which fit an empirical rule

$$R = R_o A^{1/3}, \qquad (2.1)$$

where R_o is the basic nuclear size parameter, and $A = Z + N$ is the number of nucleons, protons plus neutrons. The value of R_o is about 1 f; values 1.07 f, 1.2 f, and 1.44 f have been used for slightly different circumstances. Formally, the radius of the proton, $A = 1$, is just R_o, and that of the deuteron, $A = 2$, is 1.26 R_o.

In the most important single reaction in the sun, the joining of two protons to make a deuteron, new particles, a positron and a neutrino, are produced. Initially, an unstable ^{2}He or pp^{+2} species is briefly formed, and in a small fraction T of its decays generates the first important fusion product D = ^{2}H via

$$p + p \rightarrow D + e^+ + \nu_e. \qquad (1.10)$$

The neutrino is neutral and has nearly zero mass, and is hard to detect because it rarely interacts with matter. In spite of this difficulty, the neutrino flux from the sun at the earth has been measured and is now in agreement with solar models. The

Nanophysics of Solar and Renewable Energy, First Edition. Edward L. Wolf.
© 2012 Wiley-VCH Verlag GmbH & Co. KGaA. Published 2012 by Wiley-VCH Verlag GmbH & Co. KGaA.

neutrino flux is conceptually an easy way to measure the rate of ^{4}He formation in the sun because precisely two neutrinos are formed for each helium. The final pp cycle turns four protons into ^{4}He $+ 2e^+ + 2\nu_e$, with additional steps

$$p + {}^2H \rightarrow {}^3He + \gamma \tag{1.10a}$$

$$^3He + {}^3He \rightarrow 4He + p + p. \tag{1.10b}$$

The net energy release on the sun per cycle of these reactions is 26.2 MeV, excluding the energy that goes away with the neutrinos.

Fortunately, this complication with neutrinos is not present in reactions that are used on earth in Tokamak reactors, so these reactions are easier to start than the basic p–p reaction needed on the sun. The experts know that the complication in the p–p reaction slows down the burning of hydrogen to make helium in stars like the sun, and has the effect of greatly increasing the life of such stars. *For our purpose, we primarily need to understand in some detail the mechanism by which the charged and strongly repelling constituent particles (protons) get close enough to react* [11, 12, 18]. *This will be also our introduction to Schrodinger's equation, built upon the wave aspect of matter.*

2.1
Introduction: Protons in the Sun's Core

As was mentioned in Chapter 1, the sun is primarily composed of hydrogen and helium. These atoms are ionized in the sun to appear as protons, alpha particles (^{4}He^{2+} nuclei), and electrons, to make the system neutral.

The neutron is an unstable particle, with a lifetime of 880 s. So the sun has no free neutrons, and nuclei (which generally contain more neutrons than protons) have to be formed by fusion as we will discuss. The neutron is stable within a nucleus, but not stable as a free particle.

Small amounts of light elements, up to carbon or so, are present, but are not needed for a simple description of the energy release. This simplified picture is adequate to address the fusion process that produces the energy. The temperature ranges from about 5900 K at the outer surface to 15 million K at the core. The state of the matter is a dense ionized gas or plasma, and the density and pressure fall strongly going from the center to larger radius. The core is said to be the region out to $R_S/4$, and to produce 0.99 of the total power. At its surface, the sun faces vacuum, zero pressure, and zero density, but at the core the maximum density is estimated as 1.527×10^5 kg/m^3 [18].

The motions of the particles are prescribed primarily by the temperature and density, which vary from the surface of the sun to the interior. The properties of the dense ionized gas at the core can be estimated.

The core mass composition is 33.97% H and 64.05% ^{4}He, having masses, respectively, 1.67×10^{-27} kg and 6.68×10^{-27} kg and the core mass density is 1.527×10^5 kg/m^3. Therefore,

$$N_p = 0.3397 \times 1.527 \times 10^5 \text{ kg/m}^3 / 1.67 \times 10^{-27} \text{ kg} = 3.106 \times 10^{31} \text{ m}^{-3}. \tag{2.2}$$

$$N_{He} = 0.6405 \times 1.527 \times 10^5 \text{ kg/m}^3 / 6.68 \times 10^{-27} \text{ kg} = 1.46 \times 10^{31} \text{ m}^{-3}. \tag{2.3}$$

The electron concentration, then is $N_e = (3.106 + 2(1.46)) \times 10^{31} \text{ m}^{-3} = 6.026 \times 10^{31} \text{ m}^{-3}$. (For comparison, the free electron and positive ion densities in a metal are on the order of $5 \times 10^{28} \text{ m}^{-3}$, about 1000 times smaller.) From $N_p = 3.106 \times 10^{31} \text{ m}^{-3}$, we can infer that the interproton spacing is $N_p^{-1/3} = 3.18 \times 10^{-11}$ m. Compared to the hydrogen atom radius $a_o = 0.0529$ nm $= 5.29 \times 10^{-11}$ m, this spacing is about 0.6 a_o, but on a femtometer basis it is large, 31 800 f. From an atomic point of view, the spacing less than the Bohr radius would mean that the Mott transition (Chapter 3) has occurred, electrons are free to roam away from their protons, even at low temperature. The protons, however, are a dilute system because their spacing greatly exceeds their charge radius. This means that only two-particle collisions will be at all likely to occur. (We will see in a moment that the classical approach distance at the available energy is 1113 f, too great a spacing for a nuclear reaction to occur).

We will need to estimate the total number of protons, N_{cp} in the sun's core, defined as $0 \leq r \leq R_S/4$. Since the sun is not a solid but a dense gas, its density strongly varies with radius. It is reported (www.nasa.gov/worldbook/sun_worldbook.html dated 11/27/2007.) that the density at $R_S/4$ is 20 g/cc, about 0.133 relative to the density at $r=0$. It is also reported [19] that the density $\varrho(r)$ decays exponentially as $\varrho(r) = \exp(-\alpha r)$. With radius in units of R_s we have $\varrho(0.25)/\varrho(0) = 0.133 = \exp(-0.25\,\alpha)$ that gives $\alpha = 8.06\ R_S^{-1}$. This function will apply to the proton density, N_p, with value at $r=0$, $N_p(0) = 3.106 \times 10^{31} \text{ m}^{-3}$.

Using this information, we can write

$$N_{cp} = N_p(0) \int_0^{0.25} 4\pi r^2 \exp(-8.06r) dr = 2.51 \times 10^{56}.$$

This estimate may be high, since it is also reported [19, 20] that the *total* number of protons in the sun is 8.9×10^{56}.

Recall from Chapter 1 that the contact distance for two protons is about 2.4 f, so the protons in the sun's core are far apart from this point of view. Since the nuclear force has a range of only about 1 f, the particles have to approach each other within a few femtometers to react. We can reasonably apply the classical speed distribution function (given in Chapter 1) to the protons at the sun's core. Making use of the core temperature 15 million K, the most probable speed is $(2kT/m)^{1/2} = 0.498 \times 10^6$ m/s, and the corresponding kinetic energy is 1293 eV. Again, taking the core temperature of 15 million K, the closest approach, call it r_2, requires $k_B T = k_c e^2 / r_2 = 1293$ eV, which gives $r_2 = 1113$ f. Since this is much larger than $r = 2.4$ f, twice the charge radius of a proton, this thermally available approach distance is far too large for any reaction to occur. So the *classical particle picture is inadequate to explain the* heating of the sun. A large interproton Coulomb barrier exists and classically particles cannot cross such a barrier, fusion could not occur.

2.2
Schrodinger's Equation for the Motion of Particles

A completely different approach, *quantum physics or nanophysics*, is needed to explain the fusion events that occur in the core of the sun. In classical physics, these protons have no chance of reacting to form deuterons as must happen, for this requires a spacing about 2.4 f, while the closest approach possible at 1293 eV is 1113 f, from the equality $k_c e^2/r = 1293$ eV. The situation is as shown in Figure 1.6, where the available energy is far below the barrier height.

To repeat, the process of fusion in the sun, with proton density and temperature approximately as outlined, has led to all of the accumulated energy on earth in the form of deposits of coal, oil, and natural gas, as well as the continuing flow of 170 PW (petawatts, this is 170×10^{15} W) to the earth by direct radiation. It is also a prototype or existence proof for controlled fusion on earth. So, this is a key process, worth some thought.

The new approach needed is based on a wave aspect for all matter particles. A hint that a wave aspect for matter particles is needed to allow the particles to cross the barrier in which they are not classically allowed to exist, comes from optics.

An analogous "evanescent wave" phenomenon occurs in optics; the evanescent wave allows light to cross or "tunnel through" an air gap between two high index of refraction glass plates. In the gap region, the light intensity falls off exponentially with spacing. Light waves obey Maxwell's equations, which are second-order differential equations.

A direct verification of de Broglie's predicted wave property of matter

$$\lambda = h/p \tag{2.4}$$

(where h is Planck's constant as appeared in Equation 1.1 and $p = mv$ is the momentum of the particle) was found by Davisson and Germer, who observed magic reflection angles for a monochromatic electron beam shone on a Ni metal crystal. Their experiment verified the prediction of de Broglie, Equation 2.4.

A more familiar geometry is the two-slit experiment shown in the Figure 2.1. The condition for a constructive interference peak behind the two slits is that the path difference $d \sin \theta = n\lambda$, where $\lambda = h/p$, h is Planck's constant as appeared in Equation 1.1 and $p = mv$ is the momentum of the particle.

To return to the proton in the core of the sun, we find $\lambda = h/p = 6.6 \times 10^{-34}/(1.67 \times 10^{-27} \times 0.498 \times 10^6) = 794$ f. This is much smaller than (about 3%) of the interproton spacing 31 800 f, so we can say that the motion of the solar core protons on the whole is as free classical particles. (When they slow down approaching contact, this is no longer true, and the Schrodinger treatment is essential to understand the fusion interaction of two protons). In comparison, the electrons in a metal, even though much less dense, are completely quantum in their motion because the

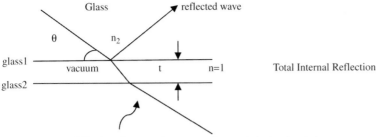

Figure 2.1 Sketch of electron diffraction in a two-slit geometry. The de Broglie condition $\lambda = h/p$ is found to predict the angles of maxima. (Courtesy of M. Medikonda).

calculated de Broglie wavelength there exceeds the interatomic spacing. This leads to a characteristic energy, the Fermi energy, which is larger than the classical energy $\frac{1}{2}kT$, with many consequences. We will need to learn about this to understand solar cells in a competent fashion.

Schrodinger found a second-order differential equation to describe matter waves. Schrodinger's equation describes in practical and accurate terms the behavior of matter particles, protons in the sun, protons in atomic nuclei, as well as electrons in atoms, metals, and semiconductors. The first appearance of the wave aspect was the relation $\lambda = h/p$, the de Broglie wavelength. Here h is Planck's constant and $p = mv$ is the classical momentum of the particle. Evaluated for the proton in the solar core, where we have found speed $v = 0.498 \times 10^6$ m/s, we find $\lambda = h/p = 6.6 \times 10^{-34}/(0.498 \times 10^6 \text{ m/s} \times 1.67 \times 10^{-27} \text{ kg}) = 794$ f. This is small compared to the interproton spacing, but large compared to the actual measured charge radius of the proton, which we have taken as 1.2 f. 794 f, the proton de Broglie wavelength, is seen also to be closer to the minimum classical spacing of two protons, 1113 f, as found for kinetic energy corresponding to 1.5×10^7 K. A second de Broglie relation gives a frequency $\omega = E/\hbar$ (here expressed in radians per second) to a particle of kinetic energy E.

The observation of electron diffraction in agreement with the de Broglie relations (see Figure 2.1) means that a matter wave described mathematically by $\Psi(x,t) = \exp(ikx - i\omega t)$ must satisfy any more general equation as was produced by Schrodinger. The wave quantity $\Psi(x,t)$ predicts the location of the particle by the relation $P(x,t) = \Psi^*(x,t)\Psi(x,t)$, where * indicates complex conjugate. Schrodinger's equation is then a statement of conservation of energy, where $k = 2\pi/\lambda$ and $\hbar = h/2\pi$.

$$[\hbar^2 k^2/2m + U - \hbar\omega]\Psi(x,t) = 0. \tag{2.5}$$

Based on this correct statement of conservation of energy, and knowing the solution $\Psi(x,t) = \exp(ikx - i\omega t)$ in case $U = 0$, the equation has to involve $\partial^2\Psi(x,t)/\partial x^2$, to generate the $\hbar^2 k^2$. In addition, the first time derivative $\partial\Psi(x,t)/\partial t = -i\omega\Psi(x,t)$ is needed in order to produce the $\hbar\omega$ in the statement of conservation of energy.

2.2.1
Time-Dependent Equation

On this basis, the Schrodinger equation in one dimension, with time-dependent potential $U(x,t)$, is

$$-(\hbar^2/2m)\partial^2\Psi(x,t)/\partial x^2 + U(x,t)\Psi(x,t) = i\hbar\partial\Psi(x,t)/\partial t = H\Psi. \tag{2.6}$$

The left-hand side of the equation is sometimes written $\mathcal{H}\Psi$, with $\mathcal{H}$ the operator that represents the energy terms, and is seen to represent the time derivative of the wavefunction.

2.2.2
Time-Independent Equation

In the common situation when the potential U is time independent, a product wave function

$$\Psi(x,t) = \psi(x)\phi(t), \tag{2.7}$$

when substituted into the time-dependent equation above, yields

$$\phi(t) = \exp(-iEt/\hbar). \tag{2.8}$$

Similarly, one obtains the *time-independent Schrodinger equation*,

$$-(\hbar^2/2m)d^2\psi(x)/dx^2 + U\psi(x) = E\psi(x), \tag{2.9}$$

to be solved for $\psi(x)$ and energy E. The solution $\psi(x)$ must satisfy the equation and also boundary conditions, as well as physical requirements.

The physical requirements are that $\psi(x)$ be continuous and have a continuous derivative except in cases where the U is infinite. $\psi(x)$ is zero where the potential U is infinite.

The second requirement is that the integral of $\psi(x)\psi^*(x)$ over the whole range of x must be finite, so that a normalization can be found.

The solutions for the equation are traveling waves when $E > U$, as would apply for protons free in the solar core, where we can take $U = 0$. On the other hand, for $E < U$, the solutions will be real exponential functions, of the form $A\exp(\kappa x) + B\exp(-\kappa x)$. In such a case, the positive exponential solution can be rejected as nonphysical. The decay constant can be seen to be

$$\kappa = [2m(V_B - E)]^{1/2}/\hbar. \tag{2.10}$$

We return to the simplest possible treatment of proton fusion in the core of the sun. Refer to Figure 1.6, from the Atzeni text on fusion. This model dates to Gamow [12] in explaining the systematics of alpha particle decay of heavy nuclei. In the decay process, the alpha particle is imagined as moving freely inside the nucleus, which is treated as a square well potential. The intermediate barrier corresponds to the Coulomb potential $k_C(Ze)(2e)/r$, where r is the spacing between the two charges.

The potential barrier has a peak at the closest spacing, $r = r_Z + r_\alpha$. In this figure, we see three regions: an inner region where bound states composed of oppositely traveling waves describe a particle bouncing around inside a fused nucleus, an intermediate range where the potential barrier exceeds the kinetic energy (the wave function will decay exponentially), and an outer region where traveling waves again exist. *These regions can be described accurately with Schrodinger's equation in spherical polar coordinates.* To simplify, the main features are usefully understood in a one-dimensional model, which we will also use for discussion of the state of electrons in a metal.

2.2.3
Bound States Inside a One-Dimensional Potential Well, E > 0

In the context of nuclear fusion, the potential U inside the nucleus will be strongly negative, on an MeV scale, compared to the potential outside the nucleus. For the purpose of simple 1D calculation, we will take the zero of energy to be that inside the nucleus, and for purpose of simple calculation, assume the potential at the outer radius L of the nucleus is infinite. In one dimension, suppose $U = 0$ for $0 < x < L$, and $U = \infty$ elsewhere, where $\psi(x) = 0$. For $0 < x < L$, the equation becomes

$$d^2\psi(x)/dx^2 + (2mE/\hbar^2)\psi(x) = 0. \tag{2.11}$$

This has the same form as the classical equation for the motion of a mass on a spring, the "simple harmonic oscillator," so the solutions can be adapted from that familiar example. (For a mass on a spring, $F = ma$, for $F = -Kx$, gives the differential equation $d^2x/dt^2 + (K/m)x = 0$, with solution $x = \sin((K/m)^{1/2}t)$. The spring constant is K, in Newtons/m. The role of K/m is taken on by $2mE/\hbar^2$ in the potential well.) Thus, one writes

$$\psi(x) = A \sin kx + B \cos kx, \tag{2.12}$$

where

$$k = (2mE/\hbar^2)^{1/2} = 2\pi/\lambda. \tag{2.13}$$

The infinite potential walls at $x = 0$ and $x = L$ require $\psi(0) = \psi(L) = 0$, which means that $B = 0$. Again, the boundary condition $\psi(L) = 0 = A \sin kL$ means that

$$kL = n\pi, \quad \text{with } n = 1, 2, \ldots \tag{2.14}$$

This, in turn, gives

$$E_n = \hbar^2(n\pi/L)^2/2m = n^2 h^2/8mL^2, \quad n = 1, 2, 3, \ldots \tag{2.15}$$

and wavefunctions, normalized to one electron per state

$$\psi_n(x) = (2/L)^{1/2} \sin(n\pi x/L). \tag{2.16}$$

We see that the allowed energies increase as the square of the integer quantum number n, and that the energies increase also quadratically as L is decreased, $E \alpha (1/L)^2$.

The condition for allowed values of $k = n\pi/L$ is equivalent to $L = n\lambda/2$, the same condition that applies to waves on a violin string.

These formulas are easily extended to the three-dimensional box of side L, by taking a product of wavefunctions (2.16) for x, y, and z, and by adding energies in Equation 2.15 according to $n^2 = n_x^2 + n_y^2 + n_z^2$.

This *exact solution* of this simple problem illustrates typical quantum behavior in which there are discrete allowed energies and corresponding wavefunctions. The wavefunctions do not precisely locate a particle, they only provide statements on the probability of finding a particle in a given range.

The conversion of the 1D picture to a *spherical finite potential well* $V(r)$ of radius a in the case of zero angular momentum gives

$$E_{n0} = \hbar^2 (n\pi/a)^2 / 2m = n^2 h^2 / 8 ma^2, n = 1, 2, 3, \ldots \tag{2.17}$$

and

$$\psi_{n00}(r) = (2\pi a)^{-1/2} [\sin(n\pi r/a)]/r. \tag{2.18}$$

The *spherical polar coordinates* are

$$x = r \sin\theta \cos\phi, \; y = r \sin\theta \sin\phi, \; z = r \cos\theta.$$

Here, we are looking at portions of the solutions that are independent of the angles, fully *spherically symmetric* solutions.

If the value of the potential at $r = a$ is a finite value V_o, then it is found that there are no bound states for

$$V_o < \hbar^2 / 8 ma^2. \tag{2.19}$$

Alternatively, we can take this as a statement that *to have a single bound state, the potential strength V_o must be at least $\hbar^2/8ma^2$.*

This is not an obvious result to obtain, but it is simple, similar to the 1D case for its energy formula, and easily remembered. The solutions for such a *finite* potential well, in either the 1D or the spherical cases, are more difficult mathematically than the infinite potential well described above. Here, the allowed quantum states (wavefunctions and energies $E < 0$) are obtained by requiring that the wavefunctions match in amplitude and in slope at $r = a$. Inside, we have $\psi_1(r) = A(\sin(kr))/r$, (Equation 2.18) and outside we have

$$\psi_2(r) = D \exp(-\kappa r)/r. \tag{2.19a}$$

Here, $k = \hbar^{-1}(2m(E + V_o))^{1/2}$ and $\kappa = \hbar^{-1}(-2mE)^{1/2}$. The solutions of a transcendental equation are needed, and must be found numerically or graphically: $-\cot z = ((z_o/z)^2 - 1)^{1/2}$, where $z_o = \hbar^{-1}(2mV_o)^{1/2}a$. There is no solution if $z_o = \hbar^{-1}(2mV_o)^{1/2}a < \pi/2$, which is equivalent to (2.19). For more details the reader may refer to the text by Griffiths [21].

This is a treatment applicable to the trapped alpha particle inside the nucleus, the left portion of Figure 1.6, an exact solution of Schrodinger's equation at the level used originally by Gamow to explain alpha particle decay!

2.3
Protons and Neutrons and Their Binding

What is a deuteron? From a simple point of view a deuteron is a proton confined in a spherical finite potential well of radius (Eq. 2.1) $a = 1.2 \, \text{f} \, 2^{1/3} = 1.51 \, \text{f}$. The potential well V_o is generated by the two-body attraction, in a useful simplification. So, we can describe the deuteron as an example of Equations 2.16–2.19. We can estimate the *minimum* barrier height as $V_o = \hbar^2/8ma^2$ taking radius $a = 1.51 \, \text{f}$. This gives

$$V_o = (6.6 \times 10^{-34}/2\pi \, 1.51 \times 10^{-15})^2 / [8 \times 1.67 \times 10^{-27} \times 1.6 \times 10^{-19}] = 2.26 \, \text{MeV}.$$

(In fact, the *binding energy* of the deuteron is known to be 2.2245 MeV, while the simple estimate we just performed would correspond to a bound state at near-zero binding energy $E_o \approx 0$. We could go through Equations 2.16–2.19 again to find V_o (a larger number would be needed) such that the bound state energy $E_o = -2.2245$ MeV. The binding energy is the result of the strong or nuclear force, whose claim on existence is indeed provided by the known nuclei, the deuteron being the smallest nucleus.

The binding energy of a nucleon when surrounded by other nucleons, say six in a cubic local environment, will be a multiple of the energy here estimated for a nucleon in contact with one other nucleon. If there are six nearest neighbors, then we would estimate $2.26 \, \text{MeV} \times 6 = 13.6 \, \text{MeV}$ per nucleon. As we will see, this is a reasonable value for the binding energy per nucleon in nuclear matter.

The proton–proton reaction that we have discussed can be put into this framework. Two protons approaching definitely experience the Coulomb barrier. The nuclear reaction is known to produce $D + e^+ + \nu_e$ suggesting that an initial doubly charged nucleus ^{2}He is formed, which in some cases decays to a deuteron, a positron, and a neutrino. This complicated decay process (in which a proton somehow mutates into a neutron, positron, and neutrino) makes the whole reaction less likely. In most cases the ^{2}He reverts to two separate protons, but the fact that this reaction occurs indicates that the Gamow tunneling process operates.

2.4
Gamow's Tunneling Model Applied to Fusion in the Sun's Core

To consider the fusion interaction of two protons, the incoming distant proton can be treated as a spherical wave $\exp(-ikr)/r$, which is valid beyond the classical turning point r_2. In the region between the outer turning point $r_2 = r_{tp}$ and the point of contact $r_1 = r_n$ (see Figure 1.6) is the *forbidden barrier region* where the solution to the Schrodinger equation, for kinetic energy less than potential energy, is a real decaying

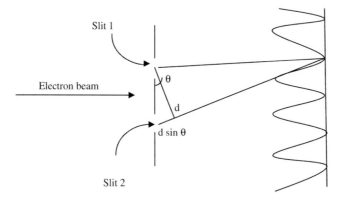

Figure 2.2 Sketch of "evanescent" light wave crossing air barrier between two glass plates. The intensity (electric field squared) of the evanescent wave decreases exponentially with air gap spacing t. This can be viewed as quantum mechanical tunneling of photons, light particles, with the role of the wave function being played by the electric field in the light wave. (Courtesy of M. Medikonda).

exponential function, quite analogous to the exponentially decaying light wave sketched in Figure 2.2.

In classical physics, the particle will never exist in this region, but nanophysics allows it in precise terms. If the barrier were constant at V_B, the wave function would be given by $\exp(-\kappa r)$, with $\kappa = (2m(V_B - E))^{1/2}/\hbar$, in the range $r_1 = r_n < r < r_2 = r_{tp}$. In this case, the tunneling probability of the particle of energy E through the barrier of height $V_B > E$ is $\exp(-2\kappa t)$, where $t = r_2 - r_1$ is the barrier thickness. The *transmission probability* is defined as

$$T = |\psi(r_1)|^2 / |\psi(r_2)|^2. \tag{2.20}$$

Here, $|\psi(r_1)|^2 = \psi^*(r_1)\psi(r_1)$. In the real case, the height of the barrier follows an $1/r$ dependence as indicated in Figure 1.6. It is difficult to solve the Schrodinger equation in case of an arbitrary barrier shape $V(r)$ and the practical approach is the simplifying WKB approximation. This useful approximation, applied to our case, gives [22]

$$T = \exp(-2\gamma). \tag{2.21}$$

With

$$\gamma = \hbar^{-1} \int_{r1}^{r2} \{2m_r[V_B(r) - E]\}^{1/2} dr. \tag{2.22}$$

Here, r_1 and r_2 are the turning points, where $E = V$, and m_r is the reduced mass

$$m_r = m_1 m_2/(m_1 + m_2). \tag{2.23}$$

It is clear in physical terms that the value of the tunneling probability T is independent of which way the particle is going. An incoming wave will be large on the outside, while an outgoing wave will be large on the inside, but the penetration probability will be the same. So the same formalism applies to our incoming wave case as to the outgoing wave case represented by the alpha particle decay mentioned above.

With $V_B(r) = k_c e^2/r$ and $m_r = m_p/2$, for two protons, with incoming energy $E = k_c e^2/r_2$ the resulting formula is

$$\gamma = \hbar^{-1}(2\,m_r E)1/2 \int_{r_1}^{r_2} (r_2/r - 1)^{1/2} dr \approx \hbar^{-1}(2\,m_r E)^{1/2}[\pi r_2/2 - 2(r_1 r_2)^{1/2}],$$

$$\text{for} \quad r_1 < r_2. \tag{2.24}$$

(This formula is reached by substituting $r = r_2 \sin^2 u$ in $\int_{r_2}^{r_2}(r_2/r-1)^{1/2}\,dr$ to get $(r_2(\pi/2 - \sin^{-1}(r_1/r_2)^{1/2}) - (r_1(r_2 - r_1))^{1/2})$. For $r_1 < r_2$, using the small-angle formula $\sin x \approx x$, one gets (2.24) [22]).

In this expression for $\gamma \approx \hbar^{-1}(2\,m_r E)^{1/2}(\pi r_2/2 - 2(r_1 r_2)^{1/2})$, for two protons $r_2 = k_c e^2/E$, and also that the $\pi r_2/2$ term in the square bracket dominates. Thus, γ is nearly proportional to $E^{-1/2}$, and the literature has adopted the notation $2\gamma = (E_G/E)^{1/2}$ with E_G called the Gamow energy, so the fusion rate will be expressed as $\exp(-(E_G/E)^{1/2})$.

One can easily see the factors that appear in E_G, by inspecting the formula $\gamma \approx \hbar^{-1}(2\,m_r E)^{1/2}(\pi r_2/2 - 2(r_1 r_2)^{1/2})$, noting that in general $r_2 = k_c Z_1 Z_2 e^2/E$ and neglecting the term in r_1.

Let us apply this formula to the two protons approaching and assume $E = 1.293$ keV, so $r_2 = 1113$ f. (It has been carefully explained that the correct energy is $E = kT$, in this case 1.293 keV, which corresponds to the peak in the velocity distribution in the center of mass frame of the two nucleons (Section 3.5 in Ref. [25])). The reduced mass is $m_p/2$, $r_1 = 2.4$ f. Then,

$$\gamma = \hbar^{-1}(2\,m_r E)^{1/2}[\pi r_2/2 - 2(r_1 r_2)^{1/2}]$$

$$= [2\pi/6.6 \times 10^{-34}][1.67 \times 10^{-27} \times 1293 \times 1.6 \times 10^{-19}]^{1/2} 10^{-15}$$

$$[1113\pi/2 - 2(1113 \times 2.4)^{1/2}] = 9.21. \tag{2.25}$$

So $T = \exp(-18.42) = 1.0 \times 10^{-8}$, which we may refer to as the tunneling probability or the *Gamow probability*, not to be confused with the temperature, for the p–p reaction, evaluated at $E = k_B T = 1.293$ KeV.

The rate of geometric collisions per proton can be estimated from the mean free path Λ. If we imagine one proton, Λ represents how far it will go before making a collision, of the type we have described, with another proton. The basic formula for the mean free path is

$$\Lambda = 1/(n\sigma) = 1/(n\pi r_2^2), \tag{2.26}$$

where $n = N_p$ is the number per unit volume of scattering centers (protons) and σ is an area if intercepted will lead to a geometric collision. We need a further probability T that a geometric collision actually leads to a fusion reaction (this is sometimes called the astrophysical S factor). This T is expected to be small based on the complicated nature of the p–p reaction, which needs to generate a positron and neutrino. Per proton, then, we can write the rate of collisions as

$$f_{coll} = (v/\Lambda) \qquad (2.27)$$

and the rate of fusion reactions per proton as

$$f_{fus} = TT f_{coll} = TT(v/\Lambda). \qquad (2.28)$$

We will calculate Λ assuming that the geometric collision corresponds to coming within the classical turning point r_2 of the second proton, this spacing was found to be 1113 f, and, for simplicity, we ignore the He ions. We earlier found that the density of protons in the core of the sun is $N_p = 3.106 \times 10^{31}$ m^{-3} and that the most probable velocity was 0.498×10^6 m/s at 15 million K.

So $\Lambda = 1/(n\,\sigma) = 1/n\,\pi r_2^2 = 8.28$ nm. (In terms of interparticle spacings, this is large, about 260 spacings). So, $f_{coll} = v/\Lambda = 0.6 \times 10^{14}$/s and $f_{fus} = TT f_{coll} = 1.0 \times 10^{-8} T \times 0.6 \times 10^{14}$/s $= 0.6 \times 10^6$ T/s. We can use this to estimate the power radiated by the sun, by multiplying by the number of participating protons N_{cp} in the core of the sun and by the energy release per reaction, and in this way get a value for the reaction probability T.

The number of protons is $N_{cp} = 6.87 \times 10^{56}$, taking the core as radius $r \leq R_S/4$, if we take the proton density as constant at 3.106×10^{31} m^{-3}. This number is too large, and we earlier estimated, in text following Equation 2.3, the number of protons in the core as $N_{cp} = 2.51 \times 10^{56}$, but mentioned that the number was still too large. Here, we take the effective number of participating protons as 1.58×10^{56}. Then, with energy release per fusion as 26.2 MeV, we get $P = 3.96 \times 10^{50} T$ watts. (The p–p cycle is (Equation 1.10a,b) initiated by the fusion reaction we are considering. The first reaction is quickly followed by several others with larger T (reaction probability) values, with a total release of energy in the sun of 26.2 MeV per initial fusion). By forcing our resulting power $P = 3.96 \times 10^{50} T$ watts equal to the known power 3.82×10^{26} watts, we estimate that the fusion reaction probability T as 9.6×10^{-25} (but we will adjust it to 8×10^{-24} later). So this reaction, we predict, is 23–24 orders of magnitude slower than a more straightforward reaction such as D + D and D + T.

It turns out that this estimate is a reasonably accurate prediction of T for the p–p reaction! We have made a simplified analysis; on the other hand, it has been stated that the reaction probability T, which is also referred to as the astrophysical S factor, is much smaller for the p–p reaction than for straight reactions not involving generation of neutrinos. To quote Atzeni (Section 1.3.3), "The p–p reaction involves a low probability beta-decay, resulting in a value of S about 25 orders of magnitude smaller than that of the DT (Deuterium-Tritium) reaction."

In the literature, this is described as a small "astrophysical cross-section," reflecting the nature of the nuclear reaction that is involved, an inverse beta decay, known to be a slow process. The first product, after the Gamow tunneling step, is an excited state, ^{2}He, of two protons, which must stabilize by emitting a positron and neutrino (inverse beta decay), else return to two separate protons. Note that the energy difference in the stabilization is positive energy 1.293 keV going to -2.22 MeV, the latter is the binding energy of the deuteron, see Figure 2.3.

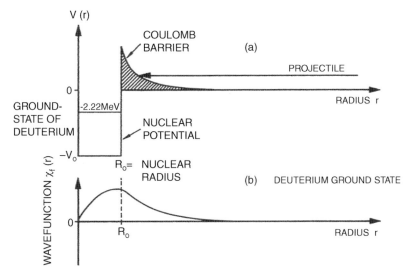

Figure 2.3 (a) Sketch of incoming proton upon Coulomb barrier, arrow terminates at classical turning point. The tunneling probability is T to reach inner radius R_o, and the further probability is TT to achieve the deuteron-bound state (b). Note that (b) is described by wavefunctions (2.18) and (1.19a), matched together at R_o. The matching in amplitude and slope, in fact, determines the bound state energy, shown as -2.22 MeV. (From [25], Figure 6.4).

A simplified view of what may happen is suggested by the known lifetime for decay of the free neutron, 880 s, into a proton, electron, and neutrino. If one assumes that the decay of the proton (into neutron, positive electron, and neutrino) has a similar time as the known decay of a neutron into a proton, electron, and neutrino, then the chance of this occurring before the excited pp state decays can be estimated as $T = \Delta t/880$, where Δt is the lifetime of the excited pp or ^{2}He state.

How can we estimate the lifetime Δt of the unstable excited state?

One estimate might be the oscillation period of the proton crossing the ^{2}He protodeuteron, 4×1.51 f$/(0.5 \times 10^6$ m/s$) = 1.21 \times 10^{-20}$ s. In this case, we get $T = 1.38 \times 10^{-23}$.

The second estimate might be from the uncertainty principle. There are two forms of the uncertainty principle, both originate in the wave aspect of particle behavior. The more familiar form is $\Delta p \, \Delta x \geq \hbar/2$, where p and x refer to the momentum and position of the particle. The less familiar form is $\Delta t \, \Delta E \geq \hbar/2$, where t and E are time and energy, respectively. We can apply the second form to estimate the lifetime as $\Delta t = \hbar/(2\Delta E)$, where we can take $\Delta E = 1.293$ keV $+ 2.22$ MeV (see Figure 2.3). In this case, we find $\Delta t = 2.95 \times 10^{-22}$ s, and probability $T = 2.95 \times 10^{-22}$ s$/880$ s $= 3.35 \times 10^{-25}$. These estimates of T are quite close to our earlier estimate $T = 9.6 \times 10^{-25}$, and the consensus from the literature as summarized by Atzeni.

So from this point of view, our simple analysis is reasonably accurate! The implication is that our simplified method might work quite well in cases where

$T = 1$ as in the case of DT (deuterium-tritium) fusion of interest for terrestrial fusion machines. The p–p reaction is really the hardest one to understand.

To return to and to emphasize our main interest in this analysis, *in classical physics the Gamow tunneling factor T would be zero, there would be no fusion!* So by explaining how the sun generates energy we have shown the necessity for Schrodinger's wave treatment of matter particles, which we will extend to atoms and solids.

You can see that we came out quite well in this simplified calculation. To compare with a more standard approach, amenable to a wide variety of fusion reactions, we mention that the major deficiency in our analysis has been to overlook the *distributions* of speeds and tunneling probabilities (cross sections) by replacing them by their most probable values. The rate of fusion is proportional to $v \times \sigma$, and in a more accurate analysis the calculated property is $\langle v\sigma \rangle$, which has units m³/s. We have seen that there is a distribution of speeds v, and the cross section will vary as the speed varies. So integrations over variables are needed. A standard framework for carrying this out gives specifically [23] for the p–p reaction

$$\langle v\sigma \rangle = 1.56 \times 10^{-43} \, T^{-2/3} \exp(-14.94/T^{1/3}) \times [1 + 0.044\, T + 2.03 \times 10^{-4}\, T^2]\, \text{m}^3/\text{s}, \quad (2.29)$$

where temperature T is expressed in keV. Evaluating this for $T = 1.293$ keV we find, for the p–p reaction at 1.5×10^7 K,

$$\langle v\sigma \rangle = 1.56 \times 10^{-43} \times 0.843 \times \exp[-14.94/1.089]$$
$$= 1.456 \times 10^{-49}[1 + .057 + .0003] = 1.54 \times 10^{-49}\, \text{m}^3/\text{s}.$$

(Note that the exponential factor here is $\exp[-14.94/1.089] = \exp[-13.71] = 1.107 \times 10^{-6}$ as compared to tunneling probability $T = 10^{-8}$ in the simplified analysis. Compared to the simplified analysis, the tunneling probability is 111 times larger, which indicates that the optimal fusion events involve higher energy particles, which, however, are fewer in number, suggested in Equations 1.8, 1.9).

We can compare this with our simplified result, by setting

$$\langle v\sigma \rangle = v\sigma. \quad (2.30)$$

Using our earlier numbers, we find, taking $\sigma = T\ T \times \pi$, $r_2^2 = 9.6 \times 10^{-25} \times 1.0 \times 10^{-8} \times \pi\, r_2^2 = 3.75 \times 10^{-56}\, \text{m}^2$ where $r_2 = 1113$ f and $v = 0.498 \times 10^6$ m/s, so

$$v\sigma = 0.498 \times 10^6\, \text{m/s}\ 3.75 \times 10^{-57}\, \text{m}^2 = 1.86 \times 10^{-50}\, \text{m}^3/\text{s}. \quad (2.31)$$

So our simplified result is only $0.121 = 1/8.3$ of the Angulo formula, setting $\langle v\sigma \rangle = v\sigma$. If we take the view that our initial approximate result is low by a factor of 8.3, it implies that the reaction constant T should be increased by the same factor, 8.3, to get $T = 8.0 \times 10^{-24}$. This value is understood as the factor by which the crucial p–p reaction is slowed down by the necessity of turning a proton into a neutron, first suggested by Bethe and Critchfield [24].

We can pause for a moment to summarize what we have learned, by looking at Figure 2.4.

2.4 Gamow's Tunneling Model Applied to Fusion in the Sun's Core

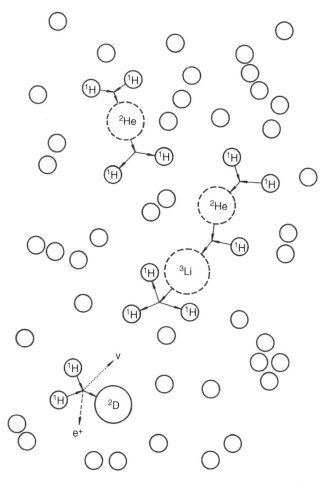

Figure 2.4 Schematic representation of hydrogen reacting with itself in the core of sun. The successful process is shown on the lower left [25].

In Figure 2.4, small circles are protons (hydrogen) denoted by turning point radius at 1293 eV, $r_2 = 1113$ f. Mean interproton spacing is 31 800 f = 49.9 r_2, in sun's core, so this picture shows more protons than a scale diagram would (at 1.5×10^7 K) by a factor of about $5^2 = 25$ if we represent the proton size by its turning point radius, 1113 f. Proton turning point collisions (e.g., lower left pair) occur for a given proton at rate $f_{coll} = v/\Lambda = 0.6 \times 10^{14}$/s, where $\Lambda = 1/(n\sigma) = 1/n\pi\, r_2^2 = 8.3$ nm $= 8.3 \times 10^6$ fm $= 7438$ $r2$. Transient protodeuteron events (one shown in upper left of the field, achieved by tunneling at Gamow probability $T = 10^{-8}$, but again releasing two protons) thus occur at rate $0.6 \times 10^6\,\mathrm{s}^{-1}$ per proton. Finally, deuteron formations (see one event in lower left portion of field, releasing one electron and one neutrino) *require further reaction probability* $T = 8.0 \times 10^{-24}$. Thus, deuteron formation, per proton, occurs at rate $f_{fus} = TT f_{coll} = 47.8 \times 10^{-19}$/s, or once every 0.67×10^{10} years. So it may be said that the lifetime of a proton (against forming a deuteron) in the sun is about 10^{10} years [26]. Yet, it is precisely these rare decays that provide the 170 petawatts heating the earth over billions of years. This is an example of a situation where rounding off 8.0×10^{-24} to zero is a serious error!

This factor $T = 8.0 \times 10^{-24}$ will not apply to reactions more likely to be used in terrestrial fusion reactors such as the D–D and D–T reactions. Fusion on earth, for this reason, should be a lot easier (by a factor of 10^{24}) than on the sun because we can start with deuterons mined from the ocean and do not have to assemble them from protons as is done on the sun. Deuterons are a necessary step on the way to making helium from hydrogen. (Again, as on the sun, there are no free neutrons that could fuse together or with protons with no Coulomb barrier, neutrons are unstable).

It is useful to consider this reaction in a broader context. In general terms, we consider a reaction of A and B to make C, in the present case A and B are both protons. In general, the rate R at which fusion product C is formed is [7]

$$R = N_A N_B (1 + \delta_{AB})^{-1} \langle v\sigma \rangle_{AB}. \tag{2.32}$$

Here $\delta_{AB} = 0$ unless A = B, in which case it equals 1.0. The units of R are $1/(m^{-3} s)$. If the energy release in the reaction is Q, the power density is $P = RQ$, which, for $N_A = N_B = N_P$, is

$$P = 0.5 \, N_p^2 \langle v\sigma \rangle Q, \tag{2.33}$$

with units watts/m^3 if Q is expressed in Joules.

We find, taking $\langle v\sigma \rangle = 1.54 \times 10^{-49}$ m^3/s, $P/m^3 = 1.54 \times 10^{-49} \times 0.5 \, (3.106 \times 10^{31})^2 \times 26.2$ MeV $\times 1.6 \times 10^{-19} = 313$ Ws/m^3 for the core of the sun. This value is quite close to a published value 276.5 W/m^3 at the center of the sun [27]. It is known that the power density falls off rapidly with increasing radius, and is 19.5 W/m^3 at 0.2 R_S. We can check this value using the total power and assuming it is generated in the core, $R \leq R_S/4$. So P/volume $= 3.82 \times 10^{26}/((4/3)\pi \, (R_S/4)^3) = 17.3$ W/m^3.

We will adopt 313 W/m^3 as a reasonable basis for scaling to a Tokamak reactor situation, in Chapter 4, using our approximate analysis summarized in Equation 2.28 by fusion rate per proton $f_{fus} = TT f_{coll} = TT (v/\Lambda)$. The working values for the center of the sun at 15 million K (1.293 KeV) are T(Gamow factor) $= 10^{-8}$, T(reaction probability in p–p reaction) $= 8 \times 10^{-24}$, thermal velocity of proton $v = 0.498 \times 10^6$ m/s.

This value, about 300 W/m^3, is a small power density, smaller than human metabolism, in agreement with other estimates. It shows that the large power output from the sun derives from its immense size. A fusion reactor on earth could be made to operate at a much higher power density, remember after all that a hydrogen bomb is a fusion reactor of a sort. For comparison, the power densities available in commercial processing tools such as *gas tungsten arc welding* (10^8 W/m^3) and *plasma torches* ($10^8 - 10^{10}$ W/m^3) are much higher [28].

We will return in Chapter 4 to analysis of a DD fusion reactor, which we can approach by adaptation of our analysis of the situation at the center of the sun.

2.5
A Survey of Nuclear Properties

The binding energy per nucleon BE has been plotted in Figure 1.5. Nuclei are characterized by a line of stability approximating $Z = N$ up to Z near 100 as shown in Figure 2.5. The stability of large nuclei disappears beyond $Z = 100$ as the Coulomb repulsion energy increases. As a function of nucleon number, A, the binding energy BE per particle rises from zero to a maximum near $A = 56$ and then gradually decreases. Nuclei beyond $A = 260$ or so are unstable.

The basic properties of the nuclei can be understood in simple terms, based upon the strong nuclear attractive force, leading to a binding energy U_o in the vicinity of 12 MeV per nucleon. Nuclei are nearly spherical, which is expected with a short-range attractive force, similar in principle to the intermolecular force in a drop of water. Water drops are spherical to minimize the number of molecules, at the surface, which will not have a full set of near neighbors. A liquid drop model was used early to understand aspects of nuclear fission. As in a water droplet, particles near the surface are less strongly bound, and this effect can be described as a surface energy.

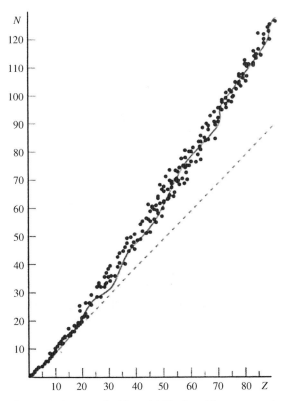

Figure 2.5 A survey of stable nuclei. The dotted line corresponds to $N = Z$ or to $A = 2Z$. Nuclei for Z much larger than 100 are unstable because of Coulomb repulsion.

It is easy to use this idea to understand the rising portion of the BE curve in Figure 1.5. If we imagine a large spherical nucleus of radius R, with A nucleons, the binding energy is $U = AU_o$ and the volume $V = 4/3\pi R^3 = (4/3)\pi R_o^3\, A$, since $R = R_o A^{1/3}$. The binding energy per unit volume is then $U_V = U_o/((4/3)\pi R_o^3)$. However, the nucleons within a distance δ of the surface, $(R-\delta) \leq r \leq R$, where δ is a measure of the range of the nuclear force, will see only *half* their binding energy, since there are no nucleons present beyond R. The "surface energy" is related to the volume $\Delta V = 4\pi R^2 \delta = 4\pi R_o^3 A^{2/3} \delta$, and the energy loss is $\tfrac{1}{2}\, U_V\, \Delta V$.

The formula for the binding energy, BE, with surface correction, is

$$\text{BE} = A\, U_o - \tfrac{1}{2} U_V \Delta V = A U_o - \tfrac{1}{2}\{U_o/[(4/3)\pi R_o^3]\} 4\pi R_o^2 A^{2/3}\delta$$
$$= A U_o[1 - 3\delta/(2 R_o A^{1/3})] \tag{2.34}$$

and the corrected BE per particle is

$$\text{BE}/A = U_o[1 - 3\delta/(2\, R_o A^{1/3})]. \tag{2.35}$$

With the empirical choices $3\delta/2R_o = 1.08$ ($\delta = 0.72 R_o$) and $U_o = 12$ MeV, the working formula

$$\text{BE}/A = 12\,\text{MeV}[1 - 1.08/A^{1/3}] \tag{2.36}$$

fits reasonably well the trend of BE values shown in Figure 1.5 for $A = 2, 3, 4$, and 56.

The second basic aspect is the Coulomb repulsion between the Z positive charges. The electrostatic repulsive energy of a uniform spherical distribution of charge $Q = Ze$ with radius R is $U_{\text{Coul}} = 3/5\, k_C\, (Ze)^2/R$.

A more subtle aspect influencing the stability of a nucleus tends to promote the nearly equal number of neutrons and protons. It is believed that this is similar to the filling of states in atoms as guided by a rule that will allow an electron of spinup and an electron of spindown in the same state, but will not allow two electrons of the same spin. In the context of nuclei, the charge, neutron versus proton, is similar to spinup or -down, and the lowest energy is achieved with similar numbers of each.

The cost of the Coulomb energy is important for large Z, which also is a reason why the proton number Z is less than the neutron number N. For large nuclei, looking at Figure 2.5, the ratio $Z/(Z+N) = Z/A$ is about 0.38.

We can incorporate these two ideas into an estimated empirical Coulomb repulsion energy per particle U_{Coul}/A (in MeV), taking $Z = 0.38\, A$, to get

$$U_{\text{Coul}}/A = (3/5\, k_C\, e^2)(0.38\, A)^2/R_o A^{4/3} = 0.104\,\text{MeV}\, A^{2/3}. \tag{2.37}$$

If we evaluate this for $A = 238$ (uranium) we get 3.99 MeV per nucleon as the Coulomb repulsion.

We can estimate the value of A at maximum in the BE curve, from the derivative of the sum of two energy terms.

$$dU/dA = d/dA\{-0.104\,\text{MeV}\, A^{2/3} + 12\,\text{MeV}\,[1 - 1.08 A^{-1/3}]\} \tag{2.38}$$

$$dU/dA = -(2/3)0.104\,\text{MeV}\, A^{-1/3} - 1/3(12\,\text{MeV})1.08\, A^{-4/3}. \tag{2.39}$$

2.5 A Survey of Nuclear Properties

Multiplying by $A^{4/3}$ and setting $dU/dA = 0$, we find $A = 62$. This is not far from the observed value, usually quoted as Fe at $A = 56$.

The total BE/A function is then approximately

$$BE/A = 12[1 - 1.08\, A^{-1/3}] - 0.104\, A^{2/3} \text{ (in MeV)}. \tag{2.40}$$

The first term is the attractive short-range force with a surface correction that roughly describes the increase in BE per nucleon as the number of nearest neighbors increases. The second term is the Coulomb repulsion that is long range and is seen to increase on a per nucleon basis as $A^{2/3}$. This function describes approximately the data curve shown in Figure 1.5.

Using this we can estimate the energy release in a hypothetical fission reaction $^{236}U \rightarrow 2\, ^{118}Pd$ using the formula (2.40). The BE for ^{236}U is 1400 MeV, while the energy for two ^{118}Pd nuclei is 1617.9 MeV. The energy release is 217.9 MeV, which is close to a typical figure quoted as 200 MeV per fission. It is clear that the primary change is in the Coulomb energy. This is not a practical reaction, although ^{236}U is a starting point reached by capture of a slow neutron by ^{235}U. The fission products are typically two nuclei of different A, for example, Kr and Ba, plus an average of 2.5 neutrons.

On the other hand, for the fusion reaction $2D \rightarrow\, ^4He$ we find from our approximate formula an energy release of 8.89 MeV. This is seriously wrong, since the quoted reaction $2D \rightarrow\, ^4He + \gamma$ lists $Q = 23.85$ MeV. It is clear also from the plot in Figure 1.5 that the ^4He or alpha particle is exceptionally stable. Shell structures in nuclei, such as this strongly bound unit of four nucleons, play a role beyond the general picture of a liquid of closely interacting nucleons.

While our understanding makes clear that the size of atomic nuclei is limited to A values less than about 240, with radius around 7.5 f, extended neutron matter is believed to exist in neutron stars. This neutron matter is apparently stabilized, in the absence of protons, by pressure.

We can find the *mass density* of nuclear matter from our working radius formula, since the mass is $Am_p = A\, 1.67 \times 10^{-27}$ kg. Setting $R = 1$ m, volume $= (4/3)\pi R^3 = (4/3)\pi\, (1.2 \times 10^{-15})^3\, A\, m^3$ with mass $A\, 1.67 \times 10^{-27}$ kg. The density is 2.307×10^{17} kg/m^3. The densities of neutron stars are estimated as two or three times this value. These values are seen to be much greater than density of the solar core, given as 1.62×10^5 kg/m^3.

Large nuclei like ^{238}U have several isotopes, corresponding to different neutron numbers, N, for the same Z. The chemical identity is controlled by Z, which sets the number of electrons that will collect around that nucleus.

This is shown by the uranium alpha decay series $^{238}U \rightarrow\, ^{236}U \rightarrow\, ^{234}U \rightarrow\, ^{232}U \rightarrow\, ^{230}U \rightarrow\, ^{228}U$, in each case by emitting an alpha particle. The decay lifetimes in this series range from about 6×10^9 years to about 300 s, all accurately predicted by the Gamow tunneling model. The plot of lifetime versus $E^{-1/2}$ gives a straight line, consistent with our earlier discussion of Equation 2.24. In the simplified expression for $\gamma \approx \hbar^{-1}\, (2m_r E)^{1/2}\, (\pi r_2/2 - 2(r_1 r_2)^{1/2})$, note that $r_2 = k_c\, 90 \times 2e^2/E$ (for uranium, $Z = 92$) and also that the $\pi r_2/2$ term in the square bracket dominates. Thus, γ is nearly proportional to $E^{-1/2}$. These considerations make clear that a plot of log of lifetime versus $1/(E)^{1/2}$ will be a straight line as shown. Similarly, the thorium series $^{232}\text{Th} \rightarrow {}^{230}\text{Th} \rightarrow {}^{228}\text{Th} \rightarrow {}^{226}\text{Th}$ is also well fit. The value $r_2 = k_c\ 88 \times 2e^2/E$ (for thorium, $Z = 90$) is slightly smaller than for uranium, leading to systematically shorter decay times.

Formally, all of these isotopes are *metastable*, trapping alpha particles at *positive* energies. In Figures 1.5 and 2.3a, these alpha particle levels would be positive, ranging from 6.8 to 4.3 MeV. In the decay process, the alpha particle is imagined as moving freely inside the nucleus, which is treated as a square well potential. The rate of decay is formally the product of a collision frequency against the wall and the Gamow tunneling probability, T. The intermediate barrier corresponds to the Coulomb potential $k_C Ze)(2e)/r$, where r is the spacing between the two charges. The peak of the Coulomb potential for ^{238}U decay would be at contact between 90 protons inside, 2 protons outside at spacing set by the nuclear radii for $A = 234$ and $A = 4$. This barrier energy is

$$V_B = k_C(90e)(2)/[R_o(41/3 + 2341/3)] = 27.8\ \text{MeV}.$$

In the Gamow tunneling probability, the reduced mass is $4 \times 234/(238) = 3.93$, which is nearly 4 (proton masses).

The data in Figure 2.6 are well fit by Gamow's application of Schrodinger's equation. These fits absolutely require quantum mechanics, which is what we are trying to establish as central to our understanding of matter on the scale of nuclei and atoms. Such fits can come only from quantum theory, there is no way for alpha decay to occur in a classical theory. The fit is remarkable in that it accurately covers decay times from a billion years to a few minutes.

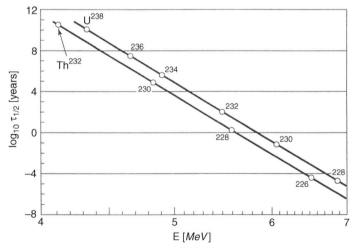

Figure 2.6 Alpha decay ($^4\text{He}^{++}$ emission) by isotopes of uranium and thorium, fit to Gamow tunneling model. Plot of logarithm of lifetime against alpha decay versus $1/E^{1/2}$ where E is the energy of the emitted alpha particle [29].

The square well potential that has been useful in explaining alpha particle decay is a simplification, we know that the only interaction is the short-range attraction between nearest-neighbor nucleons. The outer edge of the well will be rounded off again on the scale of $\delta \approx R_o$ and the physical reason for the edge of the well is simply that outside the boundary there are no more particles to be attracted to. The whole distribution is spherical in order to maximize the attractive energy, which can be restated as minimizing a (fictitious) repulsive surface energy. While the closest analogy is to a drop of water, where the intermolecular interactions are also short range, limited to nearest neighbors, the same argument, maximizing attractive (long-range) gravitational energy, explains the spherical shape of the sun. The same approach of Equation 2.35 can be applied to a metal like liquid mercury, where, however, there is no accumulating Coulomb repulsion, since each electron is accompanied by a positive ion.

It seems that the motion of the strongly bound alpha particle, within this "liquid sphere" of nucleons, is reasonably free, and this situation of free motion of a particle in a dense medium also occurs in the motion of electrons in solids such as gold or silicon. We will come back later to reasons for free motion in these cases, such free motion is beneficial to the operation of devices including solar cells. In many cases in physics, the motion of a single particle in a many-particle medium can be simplified by treating the effect of "other" particles by an effective potential or even as empty space with a boundary imposed. In physics, it is productive to use the simplest picture that works.

3
Atoms, Molecules, and Semiconductor Devices

We found in Chapter 2 that the release of energy from the sun starts when protons come together to form deuterons, a process that would not happen in classical physics. The wave property of particles of matter, as developed in the Schrodinger equation, was needed, also, to explain the alpha decay of uranium and thorium nuclei. We now extend Schrodinger's method to more familiar matter, in the form of atoms, molecules, and semiconductors. The solar cell, which produces electrical energy from sunlight, in fact, requires a sophisticated understanding of the semiconductor PN junction. So, we need to become expert in the application of Schrodinger's equation introduced in Chapter 2 to the cases of interest, including photovoltaic solar cells.

3.1
Bohr's Model of the Hydrogen Atom

To begin, we describe a useful simple model of the atom, essentially an electron orbiting around a proton. Bohr made a semiclassical model of the atom that first explained the sharply defined energy levels, a puzzling feature not present in any classical model of atoms. These levels were suggested by the optical spectra, which were composed of sharp lines. Even though this model is incorrect in some respects, it easily leads to exact results for the orbit radius, energy levels, and the wavelengths of light absorption and emission of the one electron atom. It is well worth learning.

Bohr's model describes a single electron orbiting a massive nucleus of charge $+Ze$. Bohr knew that the nucleus of the atom was a tiny object, much smaller in size than the atom itself, containing positive charge Ze, with Z the atomic number and e the electron charge, 1.6×10^{-19} C. The proton $m_p = 1.67 \times 10^{-27}$ kg is much more massive than the electron, $m_e = 9.1 \times 10^{-31}$ kg, thus $M/m = 1835$, so that nuclear motion can often be neglected. In the motion of two particles about a common center of mass, the relative motion can be corrected for the small motion of the heavier particle M by using the reduced mass $m_r = mM/(m + M)$ so $m_r \approx m_e (1 - m_e/m_p) = m_e (1 - 1/1835)$. The attractive Coulomb force $F = k_c Ze^2/r^2$, where

Nanophysics of Solar and Renewable Energy, First Edition. Edward L. Wolf.
© 2012 Wiley-VCH Verlag GmbH & Co. KGaA. Published 2012 by Wiley-VCH Verlag GmbH & Co. KGaA.

$k_c = (4\pi\varepsilon_o)^{-1} = 9 \times 10^9 \text{ Nm}^2/\text{C}^2$, must match $m_e v^2/r$, which is the mass of the electron, $m_e = 9.1 \times 10^{-31}$ kg, times the required acceleration to the center, v^2/r. The total kinetic energy of the motion, $E = mv^2/2 - k_c Ze^2/r$, adds up to $-k_c Ze^2/2r$. This is true because the kinetic energy is always -0.5 times the (negative) potential energy in a circular orbit, as can be deduced from the mentioned force balance.

There is thus a crucial relation between the total energy of the electron in the orbit, E, and the radius of the orbit, r:

$$E = -k_C Ze^2/2r. \tag{3.1}$$

This classical relation predicts collapse (of atoms, of all matter): for small r the energy is increasingly favorable (negative). So the classical electron would spiral in toward $r = 0$, giving off energy in the form of electromagnetic radiation. Fortunately, the positive value of Planck's constant, h, keeps this collapse from happening.

Bohr imposed an arbitrary quantum condition to stabilize his model of the atom. Bohr's postulate of 1913 was of the quantization of the angular momentum L of the electron of mass m circling the nucleus, in an orbit of radius r and speed v, as a multiple of Planck's constant $h = 6.6 \times 10^{-34}$ J s, divided by 2π:

$$L = mvr = n\hbar = nh/2\pi. \tag{3.2}$$

Here, n is the arbitrary integer quantum number $n = 1, 2, \ldots$. Note that Planck's constant, already described in Chapter 1, has the correct units, J s, for angular momentum. This additional constraint leads exactly to basic properties of electrons in hydrogen and similar one-electron atoms:

$$E_n = -k_c Ze^2/2r_n, \quad r_n = n^2 a_o/Z, \quad \text{where} \quad a_o = \hbar^2/mk_c e^2 = 0.053 \text{ nm}. \tag{3.3}$$

The energy of the electron in the nth orbit can thus be given as $-E_o Z^2/n^2$, $n = 1, 2, \ldots$, where

$$E_o = m_r k_c^2 e^4/2\hbar^2 = 13.6 \text{ eV}. \tag{3.4}$$

All of the previously puzzling spectroscopic observations of sharply defined light emissions and absorptions of one-electron atoms were nicely predicted by the simple quantum condition

$$h\nu = hc/\lambda = E_o(1/n_1^2 - 1/n_2^2). \tag{3.5}$$

The energy of the light is exactly the difference of the energy of two electron states, n_1 and n_2 in the atom. This was a breakthrough in the understanding of atoms, and explained sharp absorption lines as are seen in Figure 1.2. For example, the hydrogen $n = 3 - n = 2$ transition emits red light at 656 nm; this is called the Balmer line, present in light from the sun. In evaluating the wavelength in Equation 3.5 it is convenient to note that $hc = 1240$ eV nm, since $hc = 6.6 \times 10^{-34}$ J s $\times$ 3.0×10^8 m/s $(1/1.6 \times 10^{-19}$ J/eV$) = 1.2375 \times 10^{-6}$ eV m ≈ 1240 eV nm (Figure 3.1).

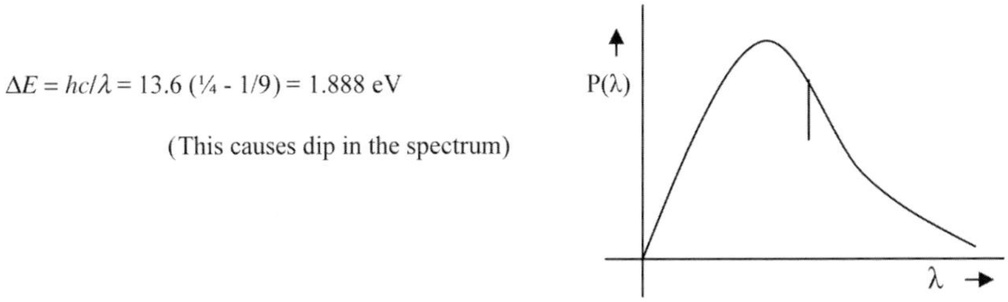

$\Delta E = hc/\lambda = 13.6\,(¼ - 1/9) = 1.888$ eV

(This causes dip in the spectrum)

Figure 3.1 Sketch of solar spectrum versus wavelength showing hydrogen $n=2$ to $n=3$ absorption at 656 nm. (Courtesy of M. Medikonda).

The wavelength of the $n=2$ to $n=3$ absorption given by $\lambda = \Delta E/hc = 1240/\Delta E = 1240/1.888 = 656$ nm is in the red part of the spectrum. As we know, the composition of the sun is 75% H by mass, so certainly hydrogen atoms are present in its atmosphere. The appearance of this sharp dip feature in the spectrum in Fig. 3.1 implies that the sun's atmosphere contains some H atoms in the $n=2$ state, with excitations ΔE of $3/4 \times 13.6$ eV $= 10.2$ eV. The probability of such an excitation of the atom is $P = \exp(-\Delta E/k_B T)$. For $T = 5973$ K, this is $P = 2.5 \times 10^{-9}$. This suggests that in the gas of hydrogen atoms above the surface of the sun, about 2.5 per billion will be able to absorb and contribute to the observed dip.

Note that the reduced mass m_r enters the energy formula Equation 3.4, and also in (3.5). In Chapter 2, we described deuterium as the first step in the energy releasing fusion process of the sun. The deuteron D forms a slightly different version of "hydrogen" atom, the only difference is the mass of the nucleus, now $2m_p$ for deuterium. We can find the small differences Δ in the energy and wavelength values for D versus H,

$$\Delta[h\nu] = \Delta[hc/\lambda] = \Delta[E_o](1/n_1^2 - 1/n_2^2), \tag{3.6}$$

which arise from the small difference in the reduced mass $\Delta m_r \approx m_e\,((1-m_e/2m_p) - (1-m_e/m_p)) = m_e\,(1 + 1/3670)$. This makes the energy for D larger by about 2.7×10^{-4} and correspondingly makes the wavelengths smaller by the same factor. For the Balmer line, 656.5 nm will be shifted by -0.178 nm to appear at 656.3 nm. This is observable, because the spectral lines are sharply defined, and is the means by which deuterium was discovered (on the sun). We will see later that in "muonic hydrogen," where the electron is replaced by a *muon*, an electron-like particle whose mass is 206.8 m_e, so that the reduced mass with the proton is $m_r \approx 186\,m_e$, that the binding energy is about $186\,E_o \approx 2528$ eV. If this "muonic" atom, furthermore, is made using deuterium, and then a molecule between two such atoms is formed, we will see later that the deuterons actually come close enough to undergo fusion at a modest rate.

The Bohr model, which does not incorporate the basic wave-like aspect of microscopic matter, fails to correctly predict some aspects of the motion and location of electrons. (It is found that the idea of an electron orbit, in the strict planetary sense, is no longer correct, in nanophysics. The uncertainty principle mentioned in Chapter 2 indicates that the position and motion of a particle cannot be known simultaneously.)

In spite of this, the Bohr model values for the electron energies $E_n = -E_o Z^2/n^2$, spectral line wavelengths, and the characteristic atomic size, $a_o = \hbar^2/mk_c e^2 = 0.053$ nm, are all exactly preserved in the fully correct treatment based on nanophysics, described in the next section.

3.2
Charge Motion in Periodic Potential

Semiconductors and their electrical conduction are important in photovoltaic cells and are strongly influenced by the wave properties of electrons. Charge motion is a central issue in efficient solar cells. We discussed the idea of the mean free path Λ for protons in the sun in Chapter 2, using formula (2.23), $\Lambda = 1/(n_s \sigma_{scatt}) = 1/n_s \pi r_2^2$, for a scattering center of radius r_2 present at density n_s. This formula is useful also in connection with the electrical conductivity, for which the conventional symbol is

$$\sigma = 1/\varrho, \tag{3.7}$$

where ϱ is the resistivity, and with σ in units Siemens or $(\Omega m)^{-1}$.

The resistivity of pure metals tends to zero at low temperature, and the ratio of the resistivity at 300 K to that at 4.2 K is called the "residual resistance ratio." It can be as large as a million in a pure crystalline sample.

The large residual resistance ratio for extremely pure metals implies large mean free path Λ at low temperature. The formula for resistivity $\varrho = 1/\sigma$ can be written in several forms:

$$\varrho = 1/Ne\mu = 1/[Ne(e\tau/m)] = m_e/(Nee\tau) = m_e v_F/(Ne^2 \Lambda). \tag{3.8}$$

Here, N is the number of electrons freely moving, τ is the time between scattering events for a given electron, v_F is a characteristic electron speed, and the *mobility* μ is defined as

$$\mu = e\tau/m. \tag{3.9}$$

The mobility, whose units are m^2/Vs, provides the *drift velocity*

$$V_D = \mu E \tag{3.10}$$

of the electron in an applied electric field $E = V/L$ given in volts per meter. Looking at the final form in (3.8), we see that the only way ϱ can increase by 10^4–10^6 going from 300 to 4 K is if mean free path Λ increases, because other factors such as N and v_F are constant. The data for pure metals in Figure 3.2 show that the interaction of electrons and metal ions is surprisingly weak: the resistivity goes to zero at low temperature. Let's look to find the most important modification of free particle motion that

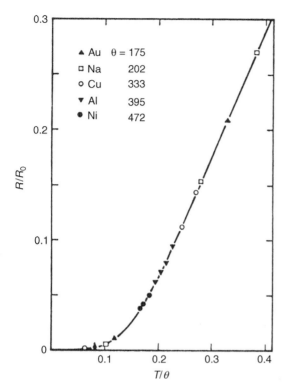

Figure 3.2 Resistivity of pure metals tends to zero at zero temperature, limited by impurities. The ratio of resistivity at 300 K to that at 4 K, called the "residual resistance ratio," can be as large as a million in an extremely pure crystalline metal. The linear region of resistivity with temperature is a well-known effect of the thermal vibrations of the atoms, characterized by θ_D, the Debye temperature. (Figure 11.5 from "Electrons and Phonons" by J. M. Ziman, Oxford University Press, 1960).

would show up if we slowly increase a weak scattering interaction between it and an array of atoms.

3.3
Energy Bands and Gaps

As we learned in connection with Figure 2.2, the Davisson–Germer experiment implies that a free electron of momentum $p = h/\lambda$ is described by a wave $\psi = e^{ikx} = e^{(i2\pi x)/\lambda}$. Suppose this wave, moving to the right, weakly scatters from atoms at spacing a, as sketched in Figure 3.3.

Here, $k = 2\pi/\lambda$ is the definition of *wave number*, the number of radians advance in phase per meter, corresponding to 2π per wavelength, λ. We see that the condition for strong coherent backreflection (*Bragg scattering*) is

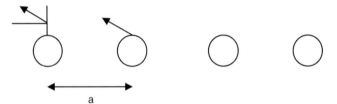

Figure 3.3 Sketch of weak scattering of an electron wave by atoms at spacing a. (Courtesy of M. Medikonda).

$$2a = n\lambda = n2\pi/k, \tag{3.11}$$

so that the wavenumber condition for backscattering is $k = n\pi/a$, where n is any integer.

Sketched in Figure 3.4 is the curve $E = \tfrac{1}{2}(\hbar^2 k^2/m)$ versus k, which can be regarded as the momentum $p/\hbar$. The curve is modified from that for a free particle, near $\pm n\pi/a$ because the scattering produces linear combinations, $\psi = e^{ikx} + e^{-ikx} = (1/2)\cos kx$ or $e^{ikx} - e^{-ikx} = (i/2)\sin kx$. These combinations are standing waves, and the energy is reduced for the $\cos kx$ versus the $\sin kx$ choice. For Si, where $a = 0.543$ nm, $k = \pi/a = \pi/0.543$ nm $= 5.78 \times 10^9$ m^{-1} and $\hbar^2 k^2/2m = 1.27$ eV, a value similar to measured values. Near $k = \pi/a$ the backscattered wave is as strong as the forward wave: the result is a standing wave (Figure 3.5). It is generally true, for Schrodinger's equation or any linear differential equation, that if one has two solutions, such as $\exp(\pm ikx)$, then linear combinations of these, such as $\sin kx$ and $\cos kx$ are equally valid, with specific choices to be based on physical reasoning. (It is also generally true that the wavefunction can be a complex quantity, such as $\psi = \exp(ikx)$, because the measured probability of finding the particle is the product of ψ with its complex conjugate, $P = \psi^*\psi$, which is a positive number.) In this case, for k just below π/a, the $\cos kx$ form is more stable, because the electron spends more time near the ions

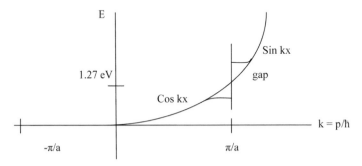

Figure 3.4 Sketch of $E(k)$ for electrons influenced by weak scattering from atoms at spacing a. The background curve corresponds to the free particle condition $E = (k\hbar)^2/2m$. The discontinuities at $k = \pm\pi/a$ arise by *coherent backscattering*, see text. The energy 1.27 eV corresponds to $(k\hbar)^2/2m$ at $k = \pm\pi/a$, using the lattice constant for crystalline silicon $a = 0.543$ nm, with $m = m_e$. This figure shows the origin of electron energy bands and gaps in a periodic lattice. (Courtesy of M. Medikonda).

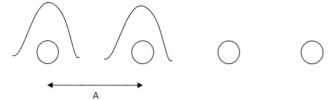

A

Figure 3.5 Sketch of standing wave probability $P = (\cos kx)^2$ expected at k just below the Bragg point π/a; the $\cos(kx)$ combination of solutions concentrates charge near the positive ions, stabilizing the state. (Courtesy of M. Medikonda).

located at $x = 0$, $x = na$. The probability density $P = (\cos kx)^2$ peaks at ion locations. For the $\sin kx$ combination, for k larger than π/a, the peaks of $P(x)$ lie between the ions.

The Kronig–Penney potential $V(x)$ model for perfect conduction along a row of atoms is sketched in Figure 3.6.

This model assumes a linear array of N atoms spaced by a along the x-axis, $0 < x < Na = L$. The 1D potential $U(x)$ is a square wave with period a:

$$U(x) = 0, \quad 0 < x < (a-w), \\ V_o, (a-w) < x < a. \tag{3.13}$$

The model potential $U(x) = 0$ except for periodic barriers of height V_o and width w. The solutions $\Psi = u_k(x) e^{ikx}$ are compatible with the 1D Schrodinger equation introduced in Chapter 2

$$(-\hbar^2/2m)d^2\Psi/dx^2 + [U(x) - E]\Psi = 0, \quad \text{with} \quad \Psi = u_k(x)e^{ikx} \tag{2.9}$$

with periodic $U(x)$ of Equation 3.13, *only* if the following condition is satisfied:

$$\cos ka = \beta(\sin qa/qa) + \cos qa = R(E), \tag{3.14}$$

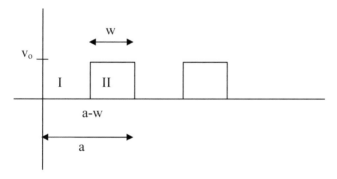

Figure 3.6 Sketch of square wave potential $U(x)$ assumed by Kronig and Penney. The band-determining condition is obtained by matching solutions of types I and II at the boundary, $x = a - w$. A useful simplification is to approximate the periodic square wave potential by repulsive delta function potentials at $x = a - w/2$, preserving the barrier potential area wV_o as $w \to 0$. (Courtesy of M. Medikonda).

where $\beta = V_o wma/\hbar^2$ and $q = (2mE)^{1/2}/\hbar$. (This simplified form of Equation 3.14 is actually obtained in a limiting process where the potential barriers are simultaneously made higher and narrower, preserving the value β. This can be described as $N\delta$ functions of strength β.) The parameter β is a dimensionless measure of the strength of the periodic variation.

We can test Equation 3.14 by examining what happens in the simple cases of vanishing and extremely strong potential barriers V.

One can see that Equation 3.14 in the zero-barrier limit, $\beta = 0$, the condition $\cos ka = \cos qa$ leads to $k = (2mE)^{1/2}/\hbar$, which recovers the free electron result, $E = \hbar^2 k^2/2m$.

Next, if β becomes arbitrarily large, the only way the term $\beta(\sin qa/qa)$ can remain finite, as the equation requires, is for $\sin qa$ to become zero. This requires $qa = n\pi$, or $a(2mE)^{1/2}/\hbar = n\pi$, which leads to $E = n^2\hbar^2/8\,ma^2$ (see Equation 2.15). These are the levels for a 1D square well of width a (recall that in the limiting process, the barrier width w goes to zero, so that each atom will occupy a potential well of width a, which is the atomic spacing).

The new interesting effects of band formation occur for finite values of β. Figure 3.7 gives a sketch of the right-hand side, $R(E)$, of Equation 3.14, versus Ka (qa in the text). Solutions of this equation are possible only when $R(E)$ is between -1 and $+1$, the range of the $\cos ka$ term on the left. Solutions for E, limited to these regions, correspond to allowed energy bands. Note that allowed solutions are possible for $-1 < \cos ka < 1$, which corresponds to $-\pi < ka < \pi$. More generally, boundaries of the allowed bands are at $k = (\pm)n\pi/a$, $n = 1, 2, 3\ldots$. It is conventional to collect the allowed bands as shown in Figure 3.8 into the range between $(\pm)\pi/a$.

The result of this analysis, then, is that in the allowed bands *no scattering occurs*, as long as the potential is periodic! This can explain why the resistivity approaches zero at low temperature for a pure crystalline metal like Au. It also explains the "band

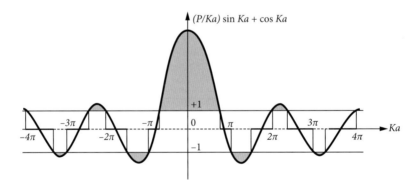

Figure 3.7 Kronig–Penney model, plot of ordinate $R(E)$ versus abscissa Ka, for $P = 4.712$. Shaded areas in this plot denote ranges of Ka that do not allow traveling wave states. Traveling wave solutions occur only when ordinate $R(E)$ has magnitude unity or less. The bottom of the conduction band E_o corresponds to $R(E) = 1$, the first crossing, and the band edge $ka = \pi$ corresponds to $R(E) = -1$. In this plot dark areas correspond to bandgaps. In the text the symbol β is used for P.

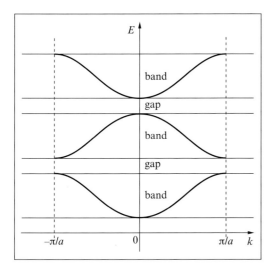

Figure 3.8 Schematic of bands E versus k in a periodic potential, based on Kronig–Penney model. The bands are restricted in k to values less than π/a. Energy gaps occurring at $k = \pm (\pi/a)$ are also physically understood on the basis of Bragg reflections at $k = \pm (\pi/a)$. Each band accommodates exactly 2N electrons, so that one electron per atom gives a half-filled band and a metal, while two electrons per atom give a filled band, an insulator.

structure" of semiconductors that are important in their application in photovoltaic devices. This effect has an analogue in electrical circuitry, where a periodic "transmission line" of lumped inductor and capacitor elements can have a pass band and a stop band. So the Kronig–Penney model explains the observation, as in Figure 3.2, of very long mean free path for electrons in pure metals, and also in semiconductors.

Each band $(-\pi/a < k < \pi/a)$ contains N atoms and can accommodate 2N electrons. In gold, one finds a half-filled band (metal) because Au releases only one electron. On the other hand, an *even* number of electrons per atom *fills* one or more bands, leading to an *insulator* because electrons cannot respond by changing k in an E field. Si has valence 4, will have *two filled bands*, and is thus a "band insulator."

3.3.1
Properties of a Metal: Electrons in an Empty Box (I)

Some essential properties of a metal such as sodium or gold can be obtained from a simple 3D model. This idealized picture is of electrons confined in a box of side L, with infinite potential outside. More realistically, the potential barrier is the "work function" φ, with typical values between 3 and 5 eV. This is a classic case of the situation mentioned at the end of Chapter 2 where the effects of "other" particles can be summarized in a useful way by a potential or just by a confining boundary condition.

The Schrodinger equation can be solved inside the box (see Equation 2.16) because of the boundary condition $\psi = 0$ at the walls, and the solutions inside an empty cube are easier to deal with than the solutions inside an empty sphere. This $\psi=0$ boundary condition is still useful for the finite barrier provided by φ, the metallic work function, whose measured value is 4.83 eV [30] for gold.

However, as we will see, the main parameter describing a metal is E_F the *Fermi energy*, which is determined only by the density of electrons, independent of the work function value.

The wavefunction (see Equation 2.16 in Chapter 2) is easily extended to three dimensions:

$$\psi = (2/L)^{3/2} \sin(n_x\pi x/L)\sin(n_y\pi y/L)\sin(n_z\pi z/L). \tag{2.16a}$$

While these were presented as bound states, they are equally valid as linear combinations of traveling waves. This is true since $\sin kx = (e^{ikx} - e^{-ikx})/i2$, and we can consider these states to be superpositions of oppositely directed traveling waves $\psi_+ = \exp(ikx)$ and $\psi_- = \exp(-ikx)$. Here, the moving waves, $\psi_\pm$, are more fundamental to a description of conduction processes.

We simulate a metal by adding electrons into the states defined by (2.16a). The important quantum aspect of this situation is Pauli's exclusion principle, such that only one electron of specified spin can occupy a state. For a given choice of n_x, n_y, and n_z only two electrons, one of spinup and one spindown, can be accommodated. If we add a large number of electrons to the box, the quantum numbers and energies of the successively filled states will be given by

$$E_n = [h^2/8m_eL^2](n_x^2 + n_y^2 + n_z^2). \tag{2.15a}$$

We need to know how the highest filled energy changes as we add electrons. To learn this, it is convenient to rewrite Equation 2.15a as

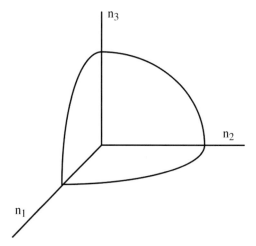

Figure 3.9 Positive octant showing spherical surface of constant energy. The number of states is twice the volume of this octant. (Courtesy of M. Medikonda).

3.3 Energy Bands and Gaps

$$E_n = E_o r^2, \tag{3.15}$$

where $E_o = (h^2/8mL^2)$, as an aid to counting the number of states filled up to an energy E, in connection with Figure 3.9. In coordinates labeled by integers n_x, n_y, and n_z, constant energy surfaces are spherical and two electron states occupy a unit volume. Since the states are labeled by positive integers, only one octant of a sphere is involved.

The number N of states out to radius r (energy E_n) is

$$N = (2)(1/8)(4\pi r^3/3) = \pi r^3/3 = (\pi/3)(E/E_o)^{3/2}, \tag{3.16}$$

for a box of side L. This is equivalent to

$$E_F = (h^2/8m)(3N/\pi L^3)^{2/3} = (h^2/8m)(3N/\pi V)^{2/3}. \tag{3.17}$$

Setting $N/V = n$, the number of states/m³, some algebra gives

$$dn/dE = g(E) = (3n/2)E^{1/2}E_F^{-3/2} = c' E^{1/2} \tag{3.18}$$

as the density of electron states per unit energy and per unit volume at energy E with c' a constant. This is the preferred form of the formula, and shows the characteristic dependence on $E^{1/2}$. We can use this to find the average kinetic energy at $T = 0$ as $E_{av} = n^{-1} \int_0^{E_F} E g(E) dE = 3/5\ E_F$. This is quite different from the classical value, $(3/2) k_B T$, and a changed form of the $P(V)$ gas law relation, from an "ideal gas" to a "degenerate gas" is a consequence we will return to.

The Fermi velocity v_F is defined by $mv_F^2/2 = E_F$, and T_F is defined by $kT_F = E_F$. The metallic Fermi velocity and Fermi temperature exceed their thermal counterparts. The reason for this is that the boundary conditions prescribe allowed states, limited to two electrons per state, raising the energy and velocity, as particles are added. A test is to compare the de Broglie wavelength of the electron to the interatomic spacing $n^{-1/3}$. Taking the highest energy state $E = E_F$ (for $n = 5.9 \times 10^{28}$ m^{-3}, $E_F = (h^2/8m)(3n/\pi)^{2/3} = 5.53$ eV), we find $\lambda = h/p = h/(2m_e E)^{1/2} = 6.6 \times 10^{-34}/(2 \times 9.1 \times 10^{-31} \times 5.53 \times 1.6 \times 10^{-19})^{1/2} = 0.52$ nm. The spacing between atoms is $n^{-1/3} = (5.9 \times 10^{28})^{-1/3} = 0.257$ nm. So, on this criterion, electrons at the Fermi energy in gold behave predominantly as waves rather than classical particles.

We will return to a more realistic discussion of electrical conduction in an "empty box" metal after we have presented a more accurate description of electronic shells in atoms.

As an application of this simple development, let us estimate the pressure of the dense electron gas in the core of the sun, following the discussion in Chapter 2. While we found that the protons in the sun act as classical particles, we must discover if the free electrons in the sun act as quantum particles or whether they are also classical in their behavior. If the electrons are following the quantum description, their equation of state is modified from the ideal gas form $P = RT/V$ (where R, the gas constant, is $N_A k_B$, the product of Avogadro's number and the Boltzmann constant), to

$$P = 2 NE_F/5\ V = 0.4\ n\ E_F. \tag{3.19}$$

(This formula [31] derives from two statements, $PV = 2/3\, NE_{av}$, which is a general ideal gas result, and, for the Fermi gas, $E_{av} = 3/5\, E_F$).

In the new formula, as expected, the Fermi energy, $E_F = (h^2/8m_e)(3n/\pi)^{2/3}$ (3.17) replaces the thermal energy $k_B T$.

To return to the condition of electrons in the sun's core, we recall that $n = N_p + 2N_{He} = 6.026 \times 10^{31}\, m^{-3}$ (see Equation 2.3) and we find, first, $E_F = (h^2/8m_e)(3n/\pi)^{2/3} = ((6.6 \times 10^{-34})^2/8m)(3 \times 6.026 \times 10^{31}/\pi)^{2/3} = 8.90 \times 10^{-17}\, J = 556\, eV$.

Since this number is less than the thermal energy $k_B T = 1293\, eV$, we can assume that the electrons at the center of the sun actually behave in a classical fashion, and formula (3.19) is not called for.

So to estimate the pressure in the sun's core, we can use the classical relation $PV = 2/3 NE_{av}$. N is the total particle density, protons, helium, and electrons, which sums to $(6.026 + 3.106 + 1.46) \times 10^{31}\, m^{-3} = 10.59 \times 10^{31}\, m^{-3}$. The pressure then is $10.59 \times 10^{31} \times 1293\, eV \times 1.6 \times 10^{-19}\, Pa = 2.19 \times 10^{16}\, Pa = 217\, Gbar$, where 1 bar $= 101\, kPa$. This is close to the value 232×10^9 bar [32] for the total hydrostatic pressure at the core of the sun. The total hydrostatic pressure is the sum of pressures from electrons, protons, He ions, and radiation. The radiation pressure is smaller, see Chapter 1 following Equation 1.2, where 0.126 Gbar was found for 15 million K. So these numbers are in good agreement, and the electrons in the sun behave classically.

To return to the properties of a system of electrons at zero temperature, states below E_F are filled and states above E_F are empty. At nonzero temperatures, the occupation (probability that the state contains one electron) is given by the Fermi function

$$f_{FD} = [\exp(\{E - E_F\}/k_B T) + 1]^{-1}. \tag{3.20}$$

The energy width of transition of f_{FD} from 1 to 0 is about $2k_B T$.

Some of these features are sketched in Figure 3.10, for the three-dimensional case, using notation $n(E) = f_{FD}(E)$ and $N(E) = g(E)$, note the characteristic $E^{1/2}$ of the upper two curves.

3.4
Atoms, Molecules, and the Covalent Bond

The Schrodinger equation introduced in Chapter 2, together with the laws of electricity and magnetism, are capable of describing details of atoms, molecules, and solids, and their interaction with photons. We need to better understand these essential methods, to allow us to extend the approach to semiconductors, PN junctions, and solar cells. We have seen that the Bohr model of the atom gives the correct energies, and allows an initial understanding of the magnetic and optical properties of one-electron atoms. But a more thorough approach, available through Schrodinger's equation in spherical polar coordinates, is necessary to incorporate the wave aspects of the electron, and to understand the nature of covalent bonding.

The atom is basically spherical, since the potential energy U of the electron in the electric field of the nucleus depends only on the radius r. The Schrodinger

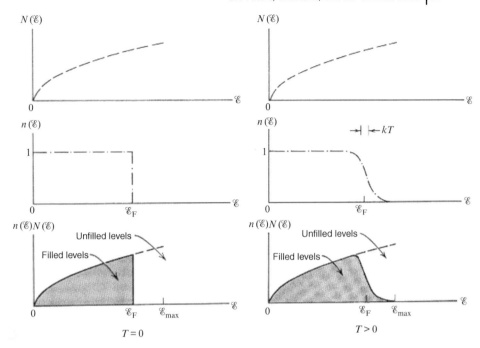

Figure 3.10 Density of states g(E) and occupation f(E) at $T=0$ (left) and T nonzero (right) in 3D case.

Equation 2.9 is easily extended to 3D x-, y-, z-coordinates (see Equation 2.16a in connection with solutions in the box of side L) but is more complicated when expressed in spherical polar coordinates (2.19). Using these coordinates, for a spherically symmetric potential $U(r)$, one finds, where θ and φ, respectively, are the polar and azimuthal angles:

$$\frac{-\hbar^2}{2m}\frac{1}{r^2}\frac{\partial}{\partial r}\left(r^2\frac{\partial \psi}{\partial r}\right)\frac{\hbar^2}{2mr^2}\left[\frac{1}{\sin\theta}\frac{\partial}{\partial \theta}\left(\sin\theta\frac{\partial \psi}{\partial \theta}\right)+\frac{1}{\sin^2\theta}\frac{\partial^2 \psi}{\partial \varphi^2}\right]+U(r)^\psi=E^\psi. \tag{3.21}$$

The Schrodinger equation is applied to the hydrogen atom, and any one-electron atom with nuclear charge Z, by choosing $U = -k_C Ze^2/r$, where k_C is the Coulomb constant. It is found, because of the spherical symmetry, that the equation separates into three equations, in variables r, θ, and φ, by setting

$$\psi = R(r)f(\theta)g(\varphi). \tag{3.22}$$

The solutions are conventionally described as the quantum states $\Psi_{n,l,m}$, specified by quantum numbers n, l, m.

The principal quantum number n, setting the energy, is associated with the solutions for the radial wavefunction,

Table 3.1 One-electron wavefunctions in real form [33].

Wavefunction designation	Wavefunction name, real form	Equation for real form of wavefunction[a], where $\varrho = Zr/a_o$ and $C_1 = Z^{3/2}/\sqrt{\pi}$
Ψ_{100}	1s	$C_1 e^{-\varrho}$
Ψ_{200}	2s	$C_2 (2-\varrho) e^{-\varrho/2}$
$\Psi_{21,\cos\varphi}$	$2p_x$	$C_2 \varrho \sin\theta \cos\varphi\, e^{-\varrho/2}$
$\Psi_{21,\sin\varphi}$	$2p_y$	$C_2 \varrho \sin\theta \sin\varphi\, e^{-\varrho/2}$
Ψ_{210}	$2p_z$	$C_2 \varrho \cos\theta\, e^{-\varrho/2}$
Ψ_{300}	3s	$C_3 (27 - 18\varrho + 2\varrho^2) e^{-\varrho/3}$
$\Psi_{31,\cos\varphi}$	$3p_x$	$C_3 (6\varrho - \varrho^2) \sin\theta \cos\varphi\, e^{-\varrho/3}$
$\Psi_{31,\sin\varphi}$	$3p_y$	$C_3 (6\varrho - \varrho^2) \sin\theta \sin\varphi\, e^{-\varrho/3}$
Ψ_{310}	$3p_z$	$C_3 (6\varrho - \varrho^2) \cos\theta\, e^{-\varrho/3}$
Ψ_{320}	$3d_{z^2}$	$C_4 \varrho^2 (3\cos^2\theta - 1) e^{-\varrho/3}$
$\Psi_{32,\cos\varphi}$	$3d_{xz}$	$C_5 \varrho^2 \sin\theta \cos\theta \cos\varphi\, e^{-\varrho/3}$
$\Psi_{32,\sin\varphi}$	$3d_{yz}$	$C_5 \varrho^2 \sin\theta \cos\theta \sin\varphi\, e^{-\varrho/3}$
$\Psi_{32,\cos2\varphi}$	$3d_{x^2-y^2}$	$C_6 \varrho^2 \sin^2\theta \cos2\varphi\, e^{-\varrho/3}$
$\Psi_{32,\sin2\varphi}$	$3d_{xy}$	$C_6 \varrho^2 \sin^2\theta \sin2\varphi\, e^{-\varrho/3}$

a) $C_2 = C_1/4\sqrt{2}$, $C_3 = 2C_1/81\sqrt{3}$, $C_4 = C_3/2$, $C_5 = \sqrt{6}C_4$, $C_6 = C_5/2$.

$$R_{n,l}(r) = (r/a_o)^l \exp(-r/na_o) L_{n,l}(r/a_o). \quad (3.23)$$

Here, $L_{n,l}(r/a_o)$ is a Laguerre polynomial in $\varrho = r/a_o$, and the radial function has $n - l - 1$ nodes. The parameter a_o is identical to its value in the Bohr model, but it no longer signifies the exact radius of an orbit. The energies of the electron states of the one electron atom, $E_n = -Z^2 E_o/n^2$ (where $E_o = 13.6$ eV, and Z is the charge on nucleus) are unchanged from the Bohr model. The energy can still be expressed as $E_n = -kZe^2/2r_n$, where $r_n = n^2 a_o/Z$, and $a_o = 0.0529$ nm is the Bohr radius, as we found in Equations 3.1–3.4 above.

The lowest energy wavefunctions $\Psi_{n,l,m}$ of the one electron atom are listed in Table 3.1 [33]. As before, to represent the hydrogen atom, set $Z = 1$. In common chemical usage, the letters s, p, d, f correspond to angular momentum values 0, 1, 2, 3. An "s-state" is spherically symmetric and has no angular momentum.

$$\Psi_{100} = (Z^{3/2}/\sqrt{\pi}\ a^{3/2}) \exp(-Zr/a_o) \quad (3.24)$$

represents the lowest energy ground state that we earlier found had energy -13.6 eV. The probability $P(r)$ of finding the electron at a radius r is

$$P(r) = 4\pi r^2 \Psi_{100}^2, \quad (3.25)$$

which is a smooth function with a maximum at $r = a_o/Z$. This is not an orbit of radius a_o, but a spherical probability cloud in which the electron's most probable radius is a_o. There is no angular momentum in this wavefunction.

This new solution represents a correction in concept, and in some numerical values, from the results of the Bohr model. Note that Ψ_{100} is real, as opposed to complex, and therefore the electron in this state has no orbital angular momentum. Both of these features correct errors of the Bohr model.

The $n=2$ wavefunctions start with Ψ_{210}, which exhibits a node in r, but is spherically symmetric. The first anisotropic (nonspherical) wavefunctions are

$$\Psi_{21,\pm 1} = R(r)f(\theta)g(\varphi) = C_2 \varrho \sin\theta \, e^{-\varrho/2} \exp(\pm i\varphi), \qquad (3.26)$$

where $\varrho = Zr/a_o$.

These are the first two wavefunctions to exhibit orbital angular momentum, here $\pm\hbar$ along the z-axis. Generally,

$$g(\varphi) = \exp(\pm im\varphi), \qquad (3.27)$$

where m, known as the magnetic quantum number, represents the projection of the orbital angular momentum vector of the electron along the z-direction, in units of $\hbar$. The orbital angular momentum **L** of the electron motion is described by the quantum numbers l and m.

The orbital angular momentum quantum number l has a restricted range of integer values:

$$l = 0, 1, 2 \ldots, n-1. \qquad (3.28)$$

This rule confirms that the ground state, $n=1$, has zero angular momentum. In the literature the letters s, p, d, f, and g, respectively, are often used to indicate $l=0, 1, 2, 3,$ and 4. So a 2s wavefunction has $n=2$ and $l=0$, and the wavefunctions sketched in Figure 3.11 are called the 2p wavefunctions.

The allowed values of the *magnetic quantum number m* depend upon both n and l according to the scheme

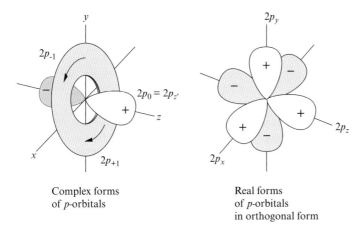

Complex forms of p-orbitals

Real forms of p-orbitals in orthogonal form

Figure 3.11 2p ($n=2, l=1$) wavefunctions in schematic form. *Left* panel, complex forms carry angular momentum. *Right* panel, linear combinations having the same energy now assume aspect of bonds.

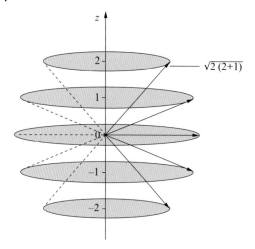

Figure 3.12 Five allowed orientations of angular momentum $l = 2$, length of vector and z projections in units of $\hbar$. Azimuthal angle φ is free to take any value.

$$m = -l, -l+1, \ldots, (l-1), l. \tag{3.29}$$

There are $2l + 1$ possibilities. Again, m represents the projection of the angular momentum vector along the z-axis, in units of $\hbar$. For $l = 1$, for example, there are three values of m, -1, 0, and 1, and this is referred to as a "triplet state." In this situation, the angular momentum vector has three distinct orientations with respect to the z-axis: polar angle $\theta = 45°$, $90°$, and $135°$. In this common notation, the $n = 2$ state (containing four distinct sets of quantum numbers) separates into a "singlet" (2s) and a "triplet" (2p).

For each electron, there is also a spin angular momentum vector **S** with length $(s(s+1))^{1/2}\hbar$, where $s = 1/2$, and projection

$$m_s = \pm(1/2)\hbar. \tag{3.30}$$

These strange rules, mathematically required to solve Schrodinger's equation, are known to accurately describe the behavior of electrons in atoms. We can use these rules to enumerate the possible distinct quantum states for a given energy state, n.

Following these rules, one can see that the number of distinct quantum states for a given n is $2n^2$ (Figure 3.12). Since the Pauli exclusion principle for electrons (and other Fermi particles) allows only one electron in each distinct quantum state, $2n^2$ is also the number of electrons that can be accommodated in the nth electron shell of an atom. For $n = 3$ this gives 18, which is seen to be twice the number of entries in Table 3.1 for $n = 3$. The factor of two represents the "spin degeneracy."

Another peculiarity of angular momentum is that the vector **L** has length $L = \sqrt{(l(l+1))}\,\hbar$ and projection $L_z = m\hbar$. A similar situation occurs for the spin vector **S**, with magnitude $S = \sqrt{(s(s+1))}\,\hbar$ and projection $m_s\hbar$. For a single electron $m_s = \pm 1/2$.

3.4 Atoms, Molecules, and the Covalent Bond

In cases where an electron has both orbital and spin angular momentum (e.g., the electron in the $n=1$ state of the one-electron atom has only S, no L), these two forms of angular momentum combine as $\mathbf{J} = \mathbf{L} + \mathbf{S}$, which has a similar rule for its magnitude: $J = \sqrt{(j(j+1))}\,\hbar$. The rules are required by the solutions of the Schrodinger equation.

The wavefunctions $\Psi_{21,\pm1} = C_2\,\varrho\sin\theta\,e^{-\varrho/2}\exp(\pm i\varphi)$ are the first two states having angular momentum. A polar plot of $\Psi_{21,\pm1}$ has a node along z, and resembles a donut flat in the x-,y-plane.

Linear combinations of states are very important in quantum mechanics. Here, the sum and difference of the $\Psi_{21,\pm1}$ states are also solutions to Schrodinger's equation, for example,

$$\Psi_{211} + \Psi_{21-1} = C_2\,\varrho\sin\theta\,e^{-\varrho/2}[\exp(i\varphi) + \exp(-i\varphi)] = C_2\,\varrho\sin\theta\,e^{-\varrho/2}\,2\cos\varphi. \tag{3.31}$$

This is twice the $2p_x$ wavefunction in Table 3.1. This linear combination, Equation 3.31, is exemplary of the real wavefunctions in Table 3.1, where linear combinations have canceled the angular momenta to provide a preferred direction for the wavefunction.

A polar plot of the $2p_x$ wavefunction (3.31) shows a node in the z-direction from the $\sin\theta$ and a maximum along the x-direction from the $\cos\varphi$, so it is a bit like a dumbbell at the origin oriented along the x-axis. Similarly, the $2p_y$ resembles a dumbbell at the origin oriented along the y-axis.

These real wavefunctions, in which the $\exp(im\varphi)$ factors have been combined to form $\sin\varphi$ and $\cos\varphi$, are more suitable for constructing bonds between atoms in molecules or in solids than are the equally valid (complex) angular momentum wavefunctions. The complex wavefunctions that carry the $\exp(im\varphi)$ factors are essential for describing orbital magnetic moments as occur in iron and similar atoms. The electrons that carry orbital magnetic moments usually lie in inner shells of their atoms.

The rules governing the one-electron atom wavefunction $\Psi_{n,l,m,m}$ and the Pauli exclusion principle, which states that only one electron can be accommodated in a completely described quantum state, are the basis for Chemical Table of the Elements. The number of electron states per atom is simply Z the nuclear charge. As we saw at the end of Chapter 2, the maximum Z for any nucleus is set by Coulomb repulsion among the Z protons.

As we have seen, the rules allow $2n^2$ distinct states for each value of the principal quantum number, n. There are several notations to describe the *filled atomic shells*. The "K shell" of an atom comprises the two electrons of $n=1$ ($1s^2$), followed by the "L shell" with $n=2$ ($2s^2 2p^6$) (Ne) and the "M shell" with $n=3$ ($3s^2 3p^6 3d^{10}$) (Ar). These closed shells contain, respectively, 2, 8, and 18 electrons. Completely filled electron cores occur at $Z=2$ (He), $Z=10$ (Ne), $Z=18$ (Ar), $Z=36$ (Kr), $Z=54$ (Xe), and $Z=86$ (Rn).

3.4.1
Properties of a Metal: Electrons in an Empty Box (II)

We come back to our description of metals, first to remark that forming the "empty box," that is, the binding of the metal atoms to make the container for the electrons, can be imagined to start with the fundamental van der Waals short-range interatomic attraction [34], but it is made stronger by an electron delocalization effect when a valence electron is weakly bound and can easily leave an atom.

Once the "empty box" is formed of real atoms, we can discuss how the electrons are bound inside it. To address the first question, we can apply Equation 2.35 to the binding of a group of metal atoms, if we replace the nuclear binding energy $U_o \rightarrow U_{coh}$, by the *cohesive energy* U_{coh} per atom for the metal. For gold U_{coh} is listed as 3.81 eV per atom, while for the rare gas atoms from Ne to Rn the cohesive energy per atom ranges from 0.02 to 0.2 eV/atom [35]. The value 3.81 eV/atom greatly exceeds the values for rare gases, which are in the range expected for the van der Waals attraction. The rare gases form liquids or weakly bound solids only at very low temperatures, but of course metals are strongly bound with high melting points. The cohesive energy is the energy needed to remove one neutral atom from the metal, which does not involve ionization of any atom.

Since the only difference between the rare gas like Radon and a monovalent metal like gold is one valence electron on the outside of a filled rare gas electron shell, this one valence electron greatly strengthens the binding.

The "surface energy" correction term (2.35) again means only that outside the metal there are no more atoms to bind a surface atom. This term describes why a liquid metal like Hg or molten solder will minimize its surface area to form a spherical drop. The collection of atoms in a metal becomes more strongly bound if the outer valence electrons "delocalize" to go into extended states similar to Equation 2.16a, which have lower kinetic energies than the bound atomic valence states.

The outer valence electron localized states are similar in concept to the hydrogenic state (Section 3.1). The valence electron for gold has principal quantum number $n = 6$, and is called a "6s" state, to indicate a spherical state of zero angular momentum. These states oscillate rapidly varying with radius (following Equation 3.23, we state that the number of radial nodes is $n - l - 1$, thus 5 for the 6s state), and the sharp $d\psi/dx$ variations lead to large kinetic energy. This part of the binding energy of a metal comes from the reduction in electron kinetic energy related to the delocalization of its wave functions. The smooth electron states extended away from the atomic cores, definitely have lower $d\psi/dx = p/\hbar$ than localized atomic states, which have rapidly varying wavefunctions (large $d\psi/dx$), as one can see by looking at Equations 2.4 and 2.9. This clearly implies a reduction of kinetic energy $p^2/2m$, and this contributes to the strength of the *metallic bond*. A more detailed estimate coming to the same conclusion is given by Kittel [36].

That the starting point for an electron in the metallic "empty box" is a free particle, rather than an atomic state, is suggested by the fact that the atomic density exceeds the Mott critical value, mentioned in Chapter 2 following Equation 2.2. Mott predicted that hydrogen atoms (see Section 3.1) when packed together at a density [37]

$$N_{Mott} \geq (0.3/a_H)^3 \qquad (3.31)$$

revert to an ionized, free electron state. While Mott did not discuss the physics in much detail, the lower kinetic energy of the delocalized electrons certainly is part of this transition (the second aspect is electron screening).

In Mott's formula a_H is the expected Bohr radius in the situation (we will see later that this value is affected by the principal quantum number n, a possible effective mass parameter and by the permittivity). For hydrogen atoms in vacuum $N_{Mott} = (0.3/0.0529 \text{ nm})^3 = 18.2 \times 10^{28} \text{ m}^{-3}$. This value is much less than the proton density in the core of the sun, $N_p = 7.14 \times 10^{31} \text{ m}^{-3}$, so that hydrogen there is certainly ionized. However, before comparing with the density of atoms in gold, 5.9×10^{28} m^{-3}, we note that the gold valence electron has quantum number $n = 6$, and therefore (see Equation 3.3) the Bohr radius is $n^2 a_o = 6^2 a_o = 1.9$ nm. So the relevant Mott concentration is reduced to $N_{Mott} \geq (0.3/(36 \times a_o))^3 = 3.91 \times 10^{24} \text{ m}^{-3}$. This allows us to proceed with our assumption that the electrons enclosed in a box of side L, imagined to be built of gold atoms, are correctly assumed to be in free particle states such as those in Equation 2.16, but more accurately, those of Equation 3.13.

We assume that the positive charge arising from the gold ions neutralizes the negative charge of the electrons, and proceed assuming the potential seen by the

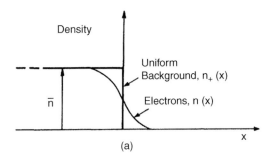

(a)

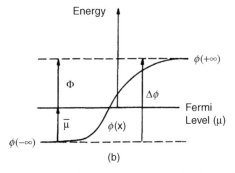

(b)

Figure 3.13 Schematic representation of (a) density distributions at a metal surface and (b) various energies relevant to a study of the workfunction [38].

electrons is $U=0$ (compared to $U=\varphi \approx \infty$ outside). In more common usage (see Figure 3.13), the work function is measured above the Fermi energy, so that the minimum energy for an electron in the interior, relative to the outside vacuum energy, is

$$U_o = -(E_F + \varphi). \tag{3.32}$$

Returning to the work function value, it is reasonable that the work function exceeds the cohesive energy U_{coh} because the former involves separating charge, ionizing an atom. In a metal, the work function barrier arises from an electric dipole layer. The electrons can tunnel slightly outside the perimeter of the metal ions, as we have discussed in connection with Gamow's probability and Figure 1.6, putting negative charge outside the metal, which will then be compensated by positive charge on inner side of the metal–vacuum boundary. This generates an electric dipole layer that leads to a jump in electric potential, the work function barrier (see Figure 3.13).

Careful calculations of the work function for a wide range of metals have been reported by Lang and Kohn [38], whose sketch of the surface dipole barrier is shown in Figure 3.10. Their theory applied to gold gives work function values 3.5, 3.65, and 3.80 eV, respectively, for 110, 100, and 111 surfaces of the face-centered cubic crystal. (The 111 surface is the body diagonal, 110 is the face diagonal, and 100 is the plane face surface, see Figure 3.14.)

We now turn to diatomic molecules, starting with H_2, but including atmospheric gases, oxygen and nitrogen. These are held together by "electron exchange," an effect that is purely quantum in its nature, although the final result is an electrostatic

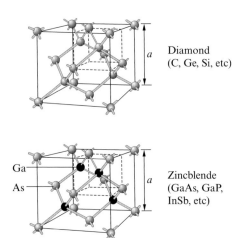

Figure 3.14 Diamond and zincblende crystal structures. Each atom is covalently bonded to four nearest neighbors in tetrahedral directions. The directed bonds are linear combinations of s and p orbitals (see Table 3.1), and analogous to directed orbitals sketched in Figure 3.1. Specifically, the linear combinations are 2s and $2p^3$ for diamond (as in CH_4) and 3s and $3p^3$ for Si. There are four valence electrons per atom, leaving a band structure with filled bands, thus insulating except for thermal excitations.

attraction. The covalent bond can involve one, two, or more electrons. The underlying effects are shown in the simplest cases, the one-electron bond in H_2^+, and the "covalent" bond in H_2. The idea of exchange or hopping between sites is brought into view by consideration of molecular bonding.

3.4.2
Hydrogen Molecule Ion H_2^+

The physics of the hydrogen molecular ion, the simplest (one-electron) covalent bond, is inherently quantum mechanical. For the system of one electron and two protons at large spacing, there are two obvious wavefunctions, $\psi_a(x_1)$ and $\psi_b(x_2)$, which represent, respectively, the electron on the left proton and on the right proton. For large spacing, these states will be long-lived, but for smaller spacing they will be unstable. An electron starting in $\psi_a(x_1)$, say, will tunnel to $\psi_b(x_2)$, at a frequency f.

To find the ground state, we make use of the idea that in quantum mechanics a linear combination of allowed solutions is also a solution. A general solution is

$$\Psi = A\psi_a(x_1) + B\psi_b(x_2), \tag{3.33}$$

where $A^2 + B^2 = 1$.

The linear combinations that are stable in time are the symmetric and antisymmetric

$$\Psi_S = 2^{-1/2}[\psi_a(x_1) + \psi_b(x_2)], \quad \Psi_A = 2^{-1/2}[\psi_a(x_1) - \psi_b(x_2)]. \tag{3.34}$$

These states are stable in time because the electron is equally present on right and left, and the tunneling instability no longer occurs.

It is easy to understand that the symmetric combination Ψ_S has a lower energy than Ψ_A, because the probability of finding the electron at the midpoint is nonzero, while that probability is zero in Ψ_A. The midpoint is an energetically favorable location for the electron because it sees attraction from both protons. The energy difference between the symmetric and the antisymmetric cases is

$$\Delta E = hf, \tag{3.35}$$

where f is the tunneling frequency of an electron started on one side for tunneling to the other side. The value ΔE is about twice the binding energy [39], which is 2.65 eV for H_2^+. So the tunneling rate is 1.28×10^{15}/s, corresponding to a residence time 0.778 fs.

The exchange of an electron between the two sites is also referred to as "hopping" or "resonance" and a more detailed treatment will give us the "hopping integral" that determines the rate. (A modification of this treatment will be applied below to a scheme for inducing D + D fusion.)

Consider two protons (at sites a and b, assume they are massive and fixed), a distance R apart, with one electron. If R is large and we can neglect interaction of the electron with the second proton, then, following the discussion of Pauling and

Wilson, we have

$$[-(\hbar^2/2m)\nabla^2 + U(r)]\psi = H\psi = E\psi \tag{3.36}$$

with r the electron position. This gives solutions, to repeat, with no interactions, $\psi = \psi_a(x_1)$ and $\psi = \psi_b(x_2)$ at energy $E = E_o$.

The interactions of the electron and the first proton with the "second" proton, $-ke^2(1/r_{a,2} + 1/R)$ are now considered. The attractive interaction, primarily occurring when the electron is between the two protons, and is attracted to both nuclear sites, stabilizes H_2^+.

We can write the interaction as

$$H_{int} = ke^2[1/R - 1/r_{a,2}], \tag{3.37}$$

where the first term is the repulsion between the two protons spaced by R. Following Pauling's treatment, one finds

$$E - E_o = (ke^2/Da_o) + (J+K)/(1+\Delta) \quad \text{for } \Psi_S \tag{3.38}$$

$$E - E_o = (ke^2/Da_o) + (J-K)/(1-\Delta), \quad \text{for } \Psi_A, \tag{3.39}$$

where

$$K = \iint \psi_b^*(x_2)[-ke^2(1/r_{a,2})]\psi_a(x_1)d^3x_1 d^3x_2 = -(ke^2/a_o)e^{-D}(1+D) \tag{3.40}$$

$$J = \iint \psi_a^*(x_1)[-ke^2(1/r_{a,2})]\psi_a(x_1)d^3x_1 d^3x_2 = (ke^2/a_o)[-D^{-1} + e^{-2D}(1+D^{-1})] \tag{3.41}$$

$$\Delta = \iint \psi_b^*(x_2)\psi_a(x_1)d^3x_1 d^3x_2 = e^{-D}(1+D+D^2/3), \quad \text{where} \quad D = R/a_o. \tag{3.42}$$

K is known as the resonance or exchange or hopping integral, and measures the rate at which an electron on one site moves to the nearest-neighbor site. One sees that its dependence on spacing is essentially $e^{-D} = e^{-R/a}$, as one would expect for a tunneling process, and that the basic energy (the prefactor of the exponential term) is $-(ke^2/a_o) = -2E_o = -27.2$ eV. In these equations, k is the Coulomb constant 9×10^9.

The energy E of the symmetric case is shown in (3.38). The major negative term is K, and this term changes sign in (3.39), the antisymmetric case. So the difference in energy between the symmetric and the antisymmetric cases is about $2K$, which amounts to about 2×2.65 eV $= 5.3$ for H_2^+. The predicted equilibrium spacing is 2.4 a_o.

The energy can be expanded as a function of $D = R/a_o$ that has a minimum at 2.4. The energy near the minimum can be expressed as

$$E(D-2.4) = E(2.4) + dE/dD(D-2.4) + \tfrac{1}{2}\,d^2E/dD^2(D-2.4)^2 + \ldots.$$
(3.38a)

Since the slope is zero at the minimum, that minimum is locally parabolic, which is the basis for simple harmonic motion, and one sees that the spring constant is $K_{spring} = d^2E/dD^2 = E''$. The oscillator frequency then is $\omega = (E''/m_{red})^{1/2}$. The value of $E''(2.4) = 0.1257\, E_o/a^2$ [40] that gives $\omega = (E''/m_{red})^{1/2} = (0.1257\ 13.6\ 1.6\ 10^{-19}/\ ^1/_2\ 1.67\ 10^{-27})^{1/2}/0.0529\,\text{nm} = 3.42 \times 10^{14}$. This corresponds to a zero point energy $\hbar\omega/2 = 0.113\,\text{eV}$.

3.5
Tetrahedral Bonding in Silicon and Related Semiconductors

The most important linear combinations of angular momentum wavefunctions (hybrids) cases are sp^3, sp^2, and sp hybrids. sp^3 hybrids describe tetrahedral bonds at 109.5° angles in methane, CH_4, and diamond. This scheme also describes bonding in the important semiconductors Si and GaAs.

In these materials, four outer electrons (two each from 3s and 3p orbitals in Si and GaAs, and two each from 4s and 4p in Ge) are stabilized into four tetrahedral covalent bond orbitals that point from each atom to its four nearest-neighbor atoms, located at apices of a tetrahedron.

The hybridization effect is essential to understanding molecules and solids. To get a better understanding, recall (see Table 3.1) that wavefunctions $\Psi_{21,\pm 1} = C_2\,\varrho \sin\theta e^{-\varrho/2}\exp(\pm i\varphi)$ are the first two states having angular momentum, where $\varrho = r/a_o$ and φ is the angle in the x-,y-plane. A polar plot of $\Psi_{21,\pm 1}$ has a node along z, and resembles a donut flat in the x-,y-plane. But these wavefunctions can be combined to make equivalent wavefunctions that point in particular directions.

The sum and difference of these states are also solutions to Schrodinger's equation, and examples are

$$\Psi_{211} + \Psi_{21-1} = C_2\varrho \sin\theta\, e^{-\varrho/2}[\exp(i\varphi) + \exp(-i\varphi)] = C_2\varrho \sin\theta\, e^{-\varrho/2} 2\cos\varphi = 2\ 2p_x,$$
(3.43)

$$\Psi_{211} - \Psi_{21-1} = C_2\varrho \sin\theta\, e^{-\varrho/2}[\exp(i\varphi) - \exp(-i\varphi)] = C_2(i)\varrho \sin\theta\, e^{-\varrho/2} 2\sin\varphi = 2\ 2p_y,$$
(3.44)

and

$$\Psi_{210} = C_2\varrho \cos\theta\, e^{-\varrho/2} = 2p_z.$$
(3.45)

These linear combinations, $2p_x$, $2p_y$, and $2p_z$, are equivalent to the original solutions that are eigenstates of angular momentum. They are also a bit like unit vectors **i**, **j**, and **k** that point along the x-, y-, and z-directions (see Table 3.1 and Figure 3.1).

These details are more important than one might expect. The same pattern, the same angular dependences for the p and d wavefunctions, are duplicated for higher values of principal quantum number n beyond $n = 2$.

3.5.1
Connection with Directed or Covalent Bonds

We can use these functions in turn to form other hybrid states. For example, the sp^3 set of wavefunctions that point to the corners of a tetrahedron occur in molecular CCl$_4$, and in crystalline germanium, silicon, and diamond.

A tetrahedron has four vertices at equal radius from the center, and this geometrical figure has both threefold (through each vertex) and twofold (through face centers) rotation axes through the origin (cube center). If we think of a cube of side $L = 2$, then four diagonally related corners, out of the eight corners of the cube, are vertices of a tetrahedron. These points have radius $\sqrt{3}$ from the center of the cube of side 2, and the spacing between adjacent vertices is $2\sqrt{2}$, which defines the bond angle β. The law of cosines ($a^2 = b^2 + c^2 - 2\,bc\cos\beta$) then gives $\cos\beta = -1/3$ or $\beta = 109.47°$ for the tetrahedral bond angle. This illustrates a method that can be applied to other bonding geometries. Accurate measurement of bond angles and bond lengths is a basic tool of the chemist and solid-state physicist.

To represent a linear combination of directed wavefunctions more simply, call it $\mathbf{r} = l\mathbf{i} + m\mathbf{j} + n\mathbf{k}$, where l, m, and n are direction cosines, which give the projection of $\mathbf{r}$ along each of the axes. To find the combinations that form the tetrahedron, imagine vectors from the center of a cube of edge 2, to four diagonally related corners, and represent these vectors by l, m, and n values. Taking the origin at the center of the cube, a suitable set of linear combinations of p_x, p_y, and p_z wavefunctions (see Table 3.1) is easily seen to be represented (by l, m, and n values) as $(1,-1,1)$, $(-1,1,1)$, $(1,1,-1)$, and $(-1,-1,-1)$.

3.5.2
Bond Angle

A useful formula for the angle Θ between two lines described, respectively, by l, m, and n values, is

$$\cos\Theta = (l_1 l_2 + m_1 m_2 + n_1 n_2)/[(l_1^2 + m_1^2 + n_1^2)^{1/2}(l_2^2 + m_2^2 + n_2^2)^{1/2}]. \quad (3.46)$$

If we apply this to find the angle, for example, between radius vectors $(1,-1,1)$ and $(-1,1,1)$, we have

$$\cos\Theta = (-1-1+1)/[(3)^{1/2}(3)^{1/2}] = -1/3. \quad \text{So} \quad \Theta = 109.47°$$

to verify the tetrahedral bonding angle we found for the diamond structures that include silicon. Again, the directed wavefunctions p_x, p_y, and p_z have been represented by unit vectors $\mathbf{i}$, $\mathbf{j}$, and $\mathbf{k}$. We have also identified a directed wavefunction with a covalent bond, which will be further discussed.

Table 3.1 includes linear combinations of 3D wavefunctions that have characteristic shapes, with lobes pointing in particular directions. These wavefunctions can also participate in hybrid bonding.

3.6
Donor and Acceptor Impurities; Charge Concentrations

On the other hand, the theory predicts that a filled band will lead to no conduction, that is, an insulator, because no empty states are available to allow motion of the electrons in response to an electric field. Such a situation is present in the pure semiconductors Si, GaAs, and Ge.

The four electrons per atom (eight electrons per cell) that fill the covalent bonds of the diamond-like structure, completely fill the lowest two "valence" bands (because of the spin degeneracy, mentioned above), leaving the 3D band empty. In concept, this would correspond to the first two bands in Figure 3.8 as being completely filled, and thus supporting no electrical conduction at zero temperature. Referring to Figure 3.8 one sees that at $k=0$, just above the second band, there is a forbidden gap, E_g. There are no states allowing conduction until the bottom of band 3, which is about 1 eV higher in these materials. For this reason, at least at low temperatures, pure samples of these materials do not conduct electricity.

Electrical conductivity at low temperature and room temperature in these materials is accomplished by "doping"; substitution for the 4-valent Si or Ge atoms either acceptor atoms of valence 3 or donor atoms of valence 5. In the case of 5-valent donor atoms like P, As, or Sb, four electrons are incorporated into tetrahedral bonding and the extra electron becomes a free electron at the bottom of the next empty band. In useful cases, the number of free electrons, n, is close to the number of donor atoms, N_D. This is termed an N-type semiconductor. In the case of boron, aluminum, and other 3-valent dopant atoms, one of the tetrahedral bonding states is unfilled, creating a "hole." A hole acts like a positive charge carrier, which moves when an electron from an adjacent bond jumps into the vacant position. Electrical conductivity by holes is dominant in a "P-type semiconductor."

The band structures for Si and GaAs are sketched in Figure 3.15. These results are calculated from approximations to Schrodinger's equation using more realistic 3D forms for the potential energy $U(x,y,z)$. The curves shown have been verified over a period of years by various experiments.

In these semiconductors, the charge carriers of importance are either electrons at a minimum in a nearly empty conduction band or holes at the top of a nearly filled valence band. In either case, the mobility $\mu = e\tau/m^*$, such that $\mathbf{v} = \mu\,\mathbf{E}$, is an important performance parameter. A high mobility is desirable as increasing the frequency response of a device such as a transistor. A useful quantity, which can be accurately predicted from the band theory, is the effective mass, m^*. This parameter is related to the inverse of the curvature of the energy band. The curvature, $\partial^2 E/\partial k^2$, can be calculated, and the formula, simply related to $E = \hbar^2 k^2/2m^*$, is

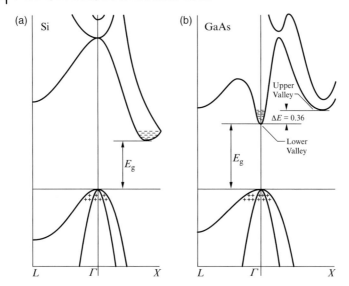

Figure 3.15 Energy band structures for silicon (a) and GaAs (b). Energy is shown vertically, measured from the top of the valence band, and k horizontally. The horizontal line marks the top of the filled "valence" bands; in pure samples, the upper bands are empty except for thermal excitations (indicated by ++ and − symbols.) The zero of momentum is indicated as "Γ", and separate sketches are given for E versus k in (111) left and (100) right directions. GaAs (b) has a "direct bandgap" because the minimum in the conduction band and the maximum in the valence band occur at the same value of wavevector k. This makes it easier for electrons and holes to recombine to produce light. The number of equivalent conduction band minima in Silicon is 6.

$$m^* = \hbar^2 / (\partial^2 E / \partial k^2). \tag{3.47}$$

While looking at the energy bands for Si and GaAs in Figure 3.15, first, one can see that generally the curvature is higher in the conduction band than in the valence band, meaning that the effective mass is smaller and the mobility, therefore, higher for electrons than for holes. Second, comparing Si and GaAs, the curvature in the conduction band minimum is higher in the latter case, leading to a higher mobility for electrons in GaAs than in Si. Another aspect is that the conduction band minima in Si are shifted from $k = 0$, which has important effects especially on the absorption of photons by Si, and similar semiconductors having an "indirect" bandgap. Parameters describing several important semiconductors are collected in Table 3.2.

The energy gaps of semiconductors range from about 0.2 eV to about 3.5 eV, as indicated in Table 3.2. Carbon in diamond form, listed at 5.5 eV, is normally considered an insulator. For devices that operate at room temperature, a gap of at least 1 eV is needed to keep the number of thermally excited carriers sufficiently low. GaN, with bandgap 3.4 eV, from a transport point of view is an insulator but from the point of view of making an injection laser is more reasonably regarded as a semiconductor. For Si with six ellipsoidal electron energy surfaces, the density of states mass is $m_{DOS}^* = (m_l^* m_t^* m_t^*)^{1/3} = 0.187$.

3.6 Donor and Acceptor Impurities; Charge Concentrations

Table 3.2 Energy gaps and other electronic parameters of important semiconductors.

Semiconductor	Bandgap (eV)		Mobility at 300 K (cm^2/V s)		Effective mass m^*/m_o		Permittivity ε
	300 K	0 K	Electrons	Holes	Electrons	Holes	
C	5.47	5.51	1800	1600	0.2	0.25	5.5
Ge	0.66	0.75	3900	1900	$m_l^* = 1.6$, $m_t^* = 0.082$	$m_{lh}^* = 0.044$, $m_{hh}^* = 0.28$	16
Si	1.12	1.16	1500	600	$m_l^* = 0.97$, $m_{DOS}^* = 0.19$	$m_{lh}^* = 0.16$, $m_{hh}^* = 0.5$	11.8
Grey Tin		~0.08					
AlSb	1.63	1.75	200	420	0.3	0.4	11
GaN	3.4				0.2	0.8	8.9
GaSb	0.67	0.80	4000	1400	0.047	0.5	15
GaAs	1.43	1.52	8500	400	0.068	0.5	10.9
GaP	2.24	2.40	110	75	0.5	0.5	10
InSb	0.16	0.26	78000	750	0.013	0.6	17
InAs	0.33	0.46	33000	460	0.02	0.41	14.5
InP	1.29	1.34	4600	150	0.07	0.4	14

The last topic relates to the origin of resistivity. Conceptually, the electron states that we have been dealing with are perfectly conducting, in the sense that an electron in such a state maintains a velocity $v = \hbar k/m$. A perfectly periodic potential gives a perfect conductor. Indeed, very pure samples of GaAs, especially epitaxial films, measured at low temperatures, give mobility values of millions, in units of cm^2/V s, and mean free paths many thousands of atomic spacings. The meaning of τ then in the expression for the mobility is the lifetime of an electron in a particular k state. The cause of limited state lifetime τ in pure metals and semiconductors at room temperature is loss of perfect periodicity as a consequence of thermal vibrations of the atoms on their lattice positions. Calculations of this effect in metals, for example, lead to the observed linear dependence of the resistivity ϱ on the absolute temperature T, as shown in Fig. 3.2.

3.6.1
Hydrogenic Donors and Excitons in Semiconductors, Direct and Indirect Bandgaps

Pure semiconductors have filled valence bands and empty conduction bands and thus have only small thermally activated electrical conductivity, depending on the size of the energy gap. As mentioned above, larger electrical conductivity is accomplished by "doping"; substitution for the 4-valent Si or Ge atoms either acceptor atoms of valence 3 or donor atoms of valence 5.

In the case of 5-valent *donor* atoms like P, As, or Sb, four electrons are incorporated into tetrahedral bonding and the extra electron, which cannot be accommodated in the already filled valence band, must occupy a state at the bottom of the next empty band, which can be close to the donor ion, in terms of its position. This free electron is attracted to the donor impurity site by the Coulomb force, and the same physics that was described for the Bohr model should apply! However, in the semiconductor medium, the Coulomb force is reduced by the relative dielectric constant, ε. Referring to Table 3.2, values of ε are large, 11.8 and 16, respectively, for Si and Ge.

The second important consideration for the motion of the electron around the donor ion is the effective mass that it acquires because of the band curvature in the semiconductor conduction band. These are again large corrections, m^*/m for electrons is about 0.2 for Si and about 0.1 for Ge. So the Bohr model can usefully be scaled by the change in dielectric constant and also by the change of electron mass to m^*.

The energy and Bohr radius are, from Equations 3.1–3.3, $E_n = -k_C Ze^2/2r_n$ and $r_n = n^2 a_o/Z$, where $a_o = \hbar^2/mk_C e^2 = 0.053$ nm. Consider the radius first, and notice that its equation contains both k_C, the Coulomb constant (proportional to $1/\varepsilon$) and the mass. So the scaled Bohr radius will be

$$a_o* = a_o(\varepsilon m/m*). \tag{3.48}$$

Similarly, considering the energy, $E_n = -k_C Ze^2/2r_n = -E_o Z^2/n^2, n = 1,2,\ldots$, where $E_o = mk_C^2 e^4/2\hbar^2$, it is evident that E scales as $m^*/(m\varepsilon^2)$:

$$E_n^* = E_n[m^*/(m\,\varepsilon^2)]. \tag{3.49}$$

For donors in Si, we find $a^* = 59 a_o = 3.13$ nm, and $E_o^* = 0.0014\,E_o = 0.0195$ eV. The large scaled Bohr radius is an indication that the continuum approximation is reasonable, and the small binding energy means that most of the electrons coming from donors in Si at room temperature escape the impurity site and move freely in the conduction band. An entirely analogous situation occurs with the holes circling acceptor sites.

An exciton is a bound state of an electron and hole created by light absorption. The important point in calculating the behavior of the exciton is to consider the reduced mass of the electron and hole as they circle about the center of mass. The exciton is an analogue of positronium, a hydrogen-like atom formed by a positron and an electron. Excitons play a role in energy transport in solar cells made of organic compounds.

3.6.2
Carrier Concentrations in Semiconductors

The objectives of this section are, first, to explain a standard method for finding N_e and N_h, the densities of conduction electrons and holes, respectively, in a semiconductor at temperature T. Second, an important special case occurs when the number of donors or acceptors becomes large and the semiconductor becomes "metallic."

At high doping, these systems behave like metals, with the Fermi energy actually lying above the conduction band edge in the N^+ case, or below the valence band edge, in the P^+ case. These separate rules apply to the so-called N^+ regions used in making contacts, and also to the "two-dimensional electron gas" (2DEG) cases that are important in making injection lasers and in forming certain "charge qubit" devices. In these "metallic" or "degenerate" cases, the number of mobile carriers remains large even at extremely low temperature, as in a metal. (In the usual semiconductor

3.6 Donor and Acceptor Impurities; Charge Concentrations

case, the carrier density goes to zero at low temperature, making the material effectively insulating.)

The essential data needed to find N_e and N_h, are

(a) the Fermi energy E_F, the energy at which the available states have 0.5 occupation probability, that is, $f_{FD}(E) = 0.5$ (see Equation 3.20);
(b) the size of the bandgap energy E_g;
(c) the temperature T; effective masses, number of equivalent band minima, and
(d) the concentrations of donor and acceptor impurities, N_D and N_A, respectively.

Smaller effects on the carrier densities comes from the values of the effective masses (Equation 3.47) for electrons and holes. For an accurate numerical work, note that for silicon, with indirect bandgap, the electron energy surfaces are ellipsoidal and the correct density of states mass is $m_{DOS}{}^* = (m_l^* m_t^* m_t^*)^{1/3} = 0.187$. For Si, with 6 equivalent band minima, the electron concentration must further be multiplied by 6.

In addition, it is important to recognize that the semiconductor as a whole must be *electrically neutral*, because each constituent atom, including the added donor and acceptor atoms, is an electrically neutral system.

To find the expected number of electrons per conduction band minimum requires integrating the density of states, multiplied by the occupation probability, over the energy range in the band. In a useful approximation, valid when the Fermi energy is several multiples of $k_B T$ below the conduction band edge, a standard result is derived, which simplifies the task of finding carrier densities N_e and N_h.

In formulas for carrier concentrations, the energy E is always measured from the top of the valence band. The standard formulas assume that the energies of electrons in the conduction band (and of holes in the valence band) vary with wavevector k as $E = \hbar^2 k^2 / 2m^*$, where $\hbar = h/2\pi$, with h Planck's constant and m^* the effective mass, as given by (3.47). In the formula for the energy E, m^* must be given in units of kg, while the common shorthand notation, as in Table 3.2, is to quote m^* as a number, in which case the reader must multiply that number by the electron mass m_e, before entering it into a calculation. (The same applies to the mass of a hole m_h^*, recalling that the motion of a hole is really the motion of an adjacent electron falling into the vacant bond position.) The bandgap, E_g, is measured from the top of the valence band $E_V = 0$, so that $E_C = E_g$.

The density of electron states (per unit energy per unit volume) in a bulk semiconductor is

$$g(E) = C_e(E - E_C)^{1/2}, \quad C_e = 4\pi(2m_e^*)^{3/2}/h^3. \tag{3.50}$$

This formula (and the similar one for the density of hole states) is closely related to (3.17), which was derived for a three-dimensional metal. Note also that the effective density of states mass $m_{DOS}{}^* = (m_l^* m_t^* m_t^*)^{1/3}$ may be appropriate.

Similarly, the density of hole states in a bulk semiconductor is given by

$$g(E) = C_h(-E)^{1/2}, \quad C_h = 4\pi(2m_e^*)^{3/2}/h^3. \tag{3.51}$$

It is evident that this formula is valid for $E < 0$, corresponding to states lying below the top of the valence band. The probability of occupation of an electron state is $f_{FD}(E)$ (3.20) and the probability that the state is occupied by a hole is $1 - f_{FD}(E)$. The shallow donor electron has a binding energy E_D. Thus, an electron at the bottom of the conduction band has energy E_g and an electron occupying a shallow donor site has energy $E = E_g - E_D$. The density of shallow donor atoms is designated N_D.

If the Fermi energy E_F is located toward the middle of the energy gap or at least a few $k_B T$ below the conduction band edge, the exponential term in the denominator of (3.20) exceeds unity, and the occupation probability is adequately given by $f_{FD}(E) \approx \exp(-(E - E_F)/k_B T)$. Thus, for the total number of electrons, we can calculate

$$N_e = \int_{E_C}^{\infty} C_e (E - E_C)^{1/2} \exp[-(E - E_F)/k_B T] dE = C_e \exp[-(E_g - E_F)/k_B T] \int_0^{\infty} x^{1/2} e^{-x} dx. \quad (3.52)$$

To get the second form of the integral change variables to $E' = E - E_C$, and then $x = E'/k_B T$, recalling that $E_C = E_g$. Move the upper limit on x to infinity, which is safe since the exponential factor falls to zero rapidly at large x, and then the value of the integral is $\pi^{1/2}/2$. Making use of the definition $C_e = 4\pi (2m_e^*)^{3/2}/h^3$, we find

$$N_C = 2(2\pi m_e^* k_B T/h^2)^{3/2}. \quad (3.53)$$

This is the effective number of states in the band at temperature T, per unit volume, for Si this must be multiplied by six to account for the six equivalent minima.

The number of electrons per unit volume in equilibrium at temperature T is then given as

$$N_e = N_C \exp\{-(E_G - E_F)/k_B T\}. \quad (3.54)$$

(It is still necessary to know E_F in order to get a numerical answer: always in such problems the central need is to find the value of E_F. In the case of a pure "intrinsic" material, where the number of electrons equals the number of holes, the Fermi level to a good approximation lies at the center of the gap, halfway between the filled and the empty states. As a first step in a pure sample, one often will assume $E_F = E_g/2$).

In the same way, the effective density of states for holes is

$$N_V = 2(2\pi m_h^* k_B T/h^2)^{3/2}. \quad (3.55)$$

The number of holes per unit volume in equilibrium at temperature T is given as

$$N_h = N_V \exp(-E_F/k_B T). \quad (3.56)$$

In a pure semiconductor, the number of holes must equal the number of electrons. By forcing the equality $N_e = N_C \exp\{-(E_G - E_F)/k_B T\} = N_h = N_V \exp\{-E_F/k_B T\}$, one can solve for E_F

$$E_F = E_g/2 + {}^3/_4 \, kT \ln(m_h^*/m_e^*) \quad \text{(for the pure sample)}. \quad (3.57)$$

Finally, consider the algebraic form of the product

$$N_e N_h = N_c N_v \exp(-E_G/k_B T). \qquad (3.58)$$

This product, which is sometimes called N_i^2 is independent of E_F, meaning that it is unchanged by doping (impurity levels)!

As an implication, if we know that there are a large number of donors N_D of small binding energy $E_D \ll k_B T$, so that it is reasonable to assume that all electrons have left the donor sites to become free electrons (i.e., $N_e = N_D$), then we find $N_h = N_i^2/N_D$, which can be small for large N_D.

In this case, since there are electrons in large numbers near E_g (at the donor level or the conduction band edge, measured from the top of the valence band), the Fermi energy (where occupation probability is $1/2$) must be near the conduction band edge.

This implies, in the heavily electron-doped case, that the probability of occupation of a hole state is extremely small, $\sim \exp(-E_G/k_B T)$. These holes (in the N-doped case) are called "minority carriers, of concentration p_n," and we see that they are small in number and have the strong temperature dependence $\exp(-E_G/k_B T)$. This is the reason for the strong temperature dependence in the reverse current of the PN junction, below.

3.6.3
The Degenerate Metallic Semiconductor

Now consider an extreme case, which will lead to an understanding of a heavily doped, metallic semiconductor, to contact solar cells, for example.

Consider pure InAs, for which Table 3.2 indicates a bandgap of 0.33 eV, electron mass $m_e^*/m_e = 0.02$, hole mass $m_h^*/m_e = 0.41$, and permittivity $\kappa = 14.5$.

First, we estimate N_e at 300 K, assuming for simplicity that the Fermi energy is at the center of the energy gap. We find $N_e = 2(2\pi m_e^* k_B T/h^2)^{3/2} \exp\{-(E_G - E_F)/k_B T\}$ $= 7.165 \times 10^{22} \times 1.7 \times 10^{-3} = 1.22 \times 10^{20}\,\text{m}^{-3}$, and the same number applies to holes for the pure sample if we neglect the shift of the Fermi energy with effective mass difference.

Next, consider heavily N-doped InAs, with $10^{18}/\text{cc} = 1.0 \times 10^{24}\,\text{m}^{-3}$ shallow donor impurities, at 300 K. This is a large doping, greatly exceeding the thermal intrinsic number of electrons, $1.22 \times 10^{20}\,\text{m}^{-3}$. What fraction of these electrons will remain on the donor sites at 300 K?

We must calculate the value of the donor binding energy, E_D, taking effective mass $m^* = 0.02$ and permittivity $\kappa = \varepsilon = 14.5$, and scale the energy from 13.6 eV for the hydrogen atom, and find 0.00129 eV. This is small compared to $k_B T$ at 300 K, which is 0.0259 eV. Thus, it is reasonable to assume that all of these electrons are in the conduction band! The same conclusion, of delocalized donor electrons, would be reached in the use of the Mott transition criterion, as expressed in Equation 3.31.

To estimate what range of energies in the conduction band will be filled by these electrons, we use the formula for the Fermi energy for a metal, (3.17), again using the effective mass $m^*/m = 0.02$, taking N/V as $1.0 \times 10^{24}\,\text{m}^{-3}$. The result is $E_F = 0.181$ eV (measured from the conduction band edge), or 0.511 eV from the valence band edge!

Formally, this would be

$$E_{FN} \approx E_G + (h^2/8m)(3N_D/\pi)^{2/3}. \tag{3.59}$$

So, heavily doped InAs is really a metal, and we would expect that the number of free electrons will not change appreciably if this sample is cooled to a low temperature.

Finally, to complete the discussion of heavily doped InAs, making use of the "mass action law" $N_e N_h = N_i^2$ at 300 K, we find that $N_h = (1.22 \times 10^{20})^2/10^{24} = 1.49 \times 10^{16}/m^3 = 1.49 \times 10^{10}$/cc. This "minority concentration" of holes is much smaller than the electron concentration, 10^{18}/cc. The minority concentration will change rapidly with varying temperature, according to $\exp(-0.33\,eV/k_B T)$.

3.7
The PN Junction, Diode I–V Characteristic, Photovoltaic Cell

To begin thinking about a PN junction in Si, we represent the semiconductor by two horizontal lines, the conduction band edge at $E = E_c$ and the valence band edge $E_V = 0$, at a distance $E_G = 1.1$ eV below. The Fermi energy is indicated by a dashed line, and for a pure Si sample it will be near the middle of the gap at 0.55 eV, as sketched in Figure 3.16a.

To introduce a PN junction, imagine at the center of our picture of the two horizontal lines, we make a transition to P-type material on the left and to N-type material on the right.

In the N-type region, using Equation 3.54, we can see that the Fermi level is given by

$$E_{FN} = E_C - k_B T \ln(N_C/N_e), \tag{3.60}$$

while on the P-type side we have

$$E_{FP} = k_B T \ln(N_V/N_h). \tag{3.61}$$

In such an imagined *abrupt junction*, electrons will quickly flow from donor levels on the right to negatively ionize acceptor levels on the left until the bands have shifted to align the Fermi level as constant across the structure. The shift will be eV_B, called the *band bending*, and V_B the *built-in potential*. The shift will be

$$E_C - k_B T[\ln(N_C/N_e) + \ln(N_V/N_h)] = eV_B. \tag{3.62}$$

This will have the effect of shifting the bands on the left upward and shifting the bands on the right downward. The voltage V_B called the built-in potential is typically near 0.6 V for Si, depending on doping levels. The width W over which the shift occurs is called the *depletion region width*:

$$W = [2\varepsilon\varepsilon_o(V_B - V)(N_D + N_A)/e(N_D N_A)]^{1/2}, \tag{3.63}$$

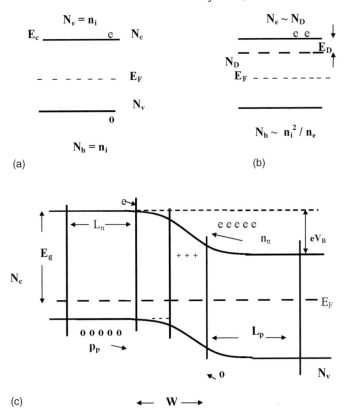

Figure 3.16 Sketches (a) intrinsic (pure) Si (b) N-type doped Si, and (c) PN junction. Electron energy vertically as in Figure 3.15, but wave vector $k=p/\hbar$, is now assumed to be fixed at the k location of the most important carriers (usually $k=0$, but displaced for electrons in Si), and x-axis denotes location in sample. It is seen (b) that Fermi energy moves from near center of gap in intrinsic material up toward conduction band edge E_c, with addition of donor impurities at concentration N_D. The concentration N_e of free electrons is given by Equation 3.54, but the zero-order estimate is $N_e \approx N_D$. (The band density of states N_C is given by Equation 3.53, similarly for N_V.) (c) shows junction formation with energy shift eV_B, with minority carriers: n_p electrons in P-region and p_n holes in N-region, which set the reverse current density. The minority electrons in diffusion length L_n (c, upper left) fall down by eV_B (across the junction) in minority carrier lifetime τ, creating reverse current density $J_{rev} \sim 2e\,(L_n n_p/\tau)$ (assuming holes behave similar to electrons). Positive applied bias voltage V (not shown) raises bands on the right, as indicated by increasing probability P for electron on the right to increase kinetic energy toward barrier height and flow to the left: $P=\exp(-e(V_B - V)/k_B T)$. Forward current in this device is electrons from right to left; forward bias reduces the band bending. (Courtesy of M. Medikonda).

where we have introduced a possible forward external bias V, seen to reduce the band shift. Note that permittivity is expressed as κ and also as ε. In practice, one side, let us assume the N-side, may be highly doped, even to produce a metallic situation as discussed before (3.59). In this case of a "one-sided junction," the width formula simplifies (for N_D large) to

$$W = [2\varepsilon\varepsilon_o(V_B - V)/eN_A]^{1/2}. \tag{3.63a}$$

In the depletion region, on the N-side, for example, the charge density is $\varrho = +N_D e$, from which, using the Poisson equation for the electric potential V, is

$$\nabla^2(V) = \varrho/\varepsilon\varepsilon_o = +N_D e/\varepsilon\varepsilon_o. \tag{3.64}$$

For one dimension, this becomes $\partial^2 V/\partial x^2 = +N_D e/\varepsilon\varepsilon_o$, so that the electric field $E = \partial V/\partial x = +N_D ex/\varepsilon\varepsilon_o + c$, with c an arbitrary constant. So the electric field increases linearly from zero, at the outer edges of the depletion layer, reaches a peak at the junction center, where $E(x)$ must be continuous from right to left. The potential $V(x)$, sketched in Figure 3.16c, is therefore quadratic in x in the depletion region, with an inflection point at the center of the junction.

We now seek to understand the outline of the PN junction in Figure 3.16c. This is the most important semiconductor device! How does it work?

It is important to understand the role of *minority carriers*, n_p electrons in P-region and p_n holes in N-region. (As we will see, these concentrations set the reverse current density.) From Equation 3.53, we see that these concentrations are small relative to the majority concentrations, which are essentially set by the levels of dopants, N_D on the N-side and N_A on the P-side. The numbers of minority carriers are set by the equilibrium equations (3.54–3.58), but the minority carrier concentration can also be understood as a balance between a generation rate (a thermal bond-breaking rate) and a *minority carrier lifetime* τ. An electron on the upper left of Figure 3.16c lives only a short-lifetime τ before it encounters one of the many holes and recombines (annihilates) either by emitting a light photon or by giving off other forms of energy. The minority electrons in diffusion length L_n (Figure 3.16c, upper left) can reach the junction electron field and cross to the other side in the minority carrier lifetime τ, so the sum of these two diffusion current densities, comprising the reverse current density, is given as

$$J_{rev} = e(L_n n_p/\tau_n) + e(L_p p_n/\tau_p). \tag{3.65}$$

Since the diffusion lengths generally are larger than W, the minority carriers are in a field-free region and diffuse randomly. Equation 3.64 can be expressed in a slightly different form by making use of the minority carrier diffusion constant D, units m²/s. D is closely related to the minority carrier mobility μ, through the relation $\mu = eD/kT$, with units expressed as (m²/V s). The diffusion length is $L = (D\tau)^{1/2}$, the distance traversed in a random walk, so the reverse current can be expressed also as

$$J_{rev} = e n_p (D_n/\tau_n)^{1/2} + e p_n (D_p/\tau_p)^{1/2} \tag{3.66}$$

and a further variation is possible using $D = k_B T\mu/e$. By using the product rule (3.58), and setting $p_p \approx N_A$, and so on, the approximate form (3.67) can be reached.

$$J_{rev} = J_o = e \cdot n_i^2 \left[\frac{D_n}{N_A L_n} + \frac{D_h}{N_D L_h} \right] \tag{3.67}$$

3.7 The PN Junction, Diode I–V Characteristic, Photovoltaic Cell

This makes explicit the strong temperature dependence of the reverse current, since we know from (3.58) that $n_i^2 = N_c N_v \exp(-E_G/k_B T)$ with strong exponential variation with T.

Since the reverse current density J_{rev} is an important parameter in the operation of solar cells, it is clear that their operating temperature is an important factor. The minority carrier diffusion length is important in solar cells. The absorbed light photons generate electron–hole pairs, and it is desirable to have the resulting minority carrier reach the junction field and transfer across to drive current in the external circuit, before recombining. In solar cell design, optimization is needed between the distance to fully absorb light (the absorption length) and the minority carrier diffusion length, to allow all of the photons to be converted, and then all of the resulting electron–hole pairs to diffuse to reach the junction field and cause external current flow.

The current density under applied bias V, from these considerations, can be expressed as

$$J = J_{rev}[\exp(eV/k_B T) - 1]. \qquad (3.68)$$

Positive applied bias voltage V (not shown in Figure 3.16c) raises bands on the right, as indicated by probability P for electron on the right to attain kinetic energy equal to barrier height and flow to the left: $P = \exp(-e(V_B - V)/k_B T)$. Forward current in this device is electrons from right to left, forward bias reduces the band bending. Positive or forward bias for this device is defined as the bias direction that reduces the band shift, and shifts the electron conduction band (on the N-side, the right side) up to allow more ready transfer of electrons into the p-type region, from right to left, the energy shift being $e(V_b - V)$. So, forward bias, positive V, reduces the shift of the bands.

At strong forward bias, with applied potential $V = V_B$, the electrons will flow from right to left without any barrier at all from the n-type side to the p-type side and this will be like a short circuit. So, the forward I–V characteristics of the PN junction are governed by the factor $\exp(e(V - V_b)/k_B T)$. It is an exponentially rising characteristic versus V.

In reverse bias, the only current that flows is the described diffusion current that is independent of applied voltage. This current comes as thermally present electrons and thermally present holes diffuse and reach the junction electric field at the center of the junction.

If light falls on the PN junction, as sketched in Figure 3.16c, the result will be generation of electron–hole pairs. This amounts to generating large numbers of minority carriers, and these photogenerated minority carriers will greatly enhance the reverse current. The resulting current density from absorbed light is described by $J = -J_L$ that would flow if the junction were short-circuited.

If the junction is open circuit, so the current is zero, then $J = J_{rev}(\exp(eV/k_B T) - 1) - J_L = 0$. That is equivalent to a maximum open-circuit voltage

$$V_{oc} = (k_B T/e)\ln(J_L/J_{rev} + 1). \qquad (3.69)$$

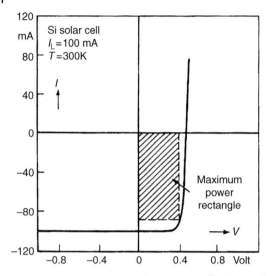

Figure 3.17 Current–voltage characteristics of a silicon solar cell under illumination [41].

If the device is terminated in a load resistor, the power density is

$$P = JV = J_{rev}V[\exp(eV/k_BT)-1] - J_LV, \qquad (3.70)$$

where the actual V will depend on the choice of the load resistance (Figure 3.17).

3.8
Metals and Plasmas

The fundamental properties of the plasma are independent motion of charges of opposite sign and charge neutrality. Each charge q is locally surrounded by its electric potential $V = k_c q/r$, but this potential will be felt by neighboring charges, which are mobile. Thus, a *shielding property* is inherent to a plasma. If a disturbance of potential, say, V appears, it will alter the local density of electrons to a perturbed value $n' = n\exp(eV/k_BT) \approx n(1 + eV/k_BT)$ assuming eV is small compared to k_BT. Recalling Equation 3.64, where we can identify the relevant charge density as $e(n'-n)$ and find, since $n' - n \approx neV/k_BT$,

$$\nabla^2(V) = n e^2 V/k_B T \varepsilon_o = V/\lambda_D^2, \qquad (3.71)$$

where

$$\lambda_D = (k_B T \varepsilon_o/ne^2)^{1/2} \qquad (3.71a)$$

is the Debye length, which is important in plasmas and also in solution chemistry where ions are mobile and can shield surface potentials, for example.

To solve (3.71) in spherical geometry as surrounding a point charge q, one finds

$$V(r) = (k_c q/r)\exp(-r/\lambda_D) \qquad (3.71b)$$

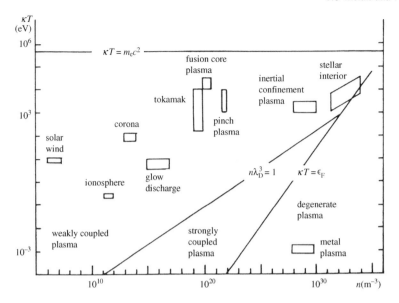

Figure 3.18 Survey of plasma domains in temperature–density representation. Note that temperatures are represented in eV units. In this figure, the "metal plasma" region might be extended to higher temperatures, since some metals, for example, W, Ta, and Hf, are definitely Fermi gases up to 1000 K = 0.08 eV. ([42] Figure 1.1).

so that a potential disturbance is shielded away in a distance λ_D. The product $n\lambda_D^3$, denoted the "plasma parameter," when large indicates that the average Coulomb energy is small compared to kinetic energy, this is seen to apply in left portions of Figure 3.18 to sun's interior and to Tokomak plasma. If $n^{-1/3}$ is r, then in this regime $\lambda_D \gg r$, so that screening does not occur. The "strongly coupled plasma" region is $n\lambda_D^3$ small, screening length short compared to mean interparticle distance, means Coulomb energy is large compared to kinetic energy, applies to metals.

4
Terrestrial Approaches to Fusion Energy

We regard deuteron fusion as a renewable energy process given the large supply of deuterons in the sea. In this chapter, we build upon what we learned about fusion in the sun with the hope of applying the same process on earth.

A summary of fusion reactions of technological interest is given in Table 4.1. We have seen in Chapter 2 that the proton–proton p–p fusion reaction that powers the sun is very difficult to achieve. The reactions of technological interest in Table 4.1 have much higher cross sections than the p–p reaction. The reactions involving D and T (tritium) that have been observed in laboratories and may power a Tokamak reactor to produce controlled fusion energy on earth are the main subject here.

In Table 4.1, Q is the kinetic energy release in the reaction, which is shared among the products depending on their masses. The cross section in units of cm^2 is that at the most favorable energy for the reaction to occur, which is given in the last column, in keV. (The typical variation of cross section with energy is sketched in Figure 4.4.)

Fusion energy can be considered renewable if based on deuterium, since the ocean contains a huge dilute reservoir of deuterons in the form of heavy water, HDO and DDO. Furthermore, the ocean contains large amounts of lithium, which can also be used as a fusion reactant [44]. The reactions shown in Table 4.1 have cross sections or probabilities about 25 orders of magnitude larger than the p–p reaction described in Chapter 2. These reactions take deuterons as starting material. An interesting question posed by the set of reactions shown is why should the D–T reaction have a higher cross section than the DD reactions? It appears that this difference is related to an aspect of the quantum mechanical process of tunneling.

The primary thrust toward producing energy from fusion is the ITER Tokamak reactor, planned with a toroidal DT plasma. ITER (http://www.iter.org/) is a $840\,m^3$ torus reactor being built in Cadamarche, France, by an international consortium. In a sense this is an attempt to scale the successful fusion conditions on the sun to parameters, notably the lower pressure, that can be attained on earth. The pressure has to be much lower, so scaling leads to a lower density of reactants, and the temperature can be raised to compensate. The temperature is easier to control in a laboratory plasma, and can exceed the temperature in the sun's core. The ITER approach is based upon magnetic confinement, to keep the 100 million K plasma from being cooled by, and damaging, the containing walls. Before we discuss the

Nanophysics of Solar and Renewable Energy, First Edition. Edward L. Wolf.
© 2012 Wiley-VCH Verlag GmbH & Co. KGaA. Published 2012 by Wiley-VCH Verlag GmbH & Co. KGaA.

4 Terrestrial Approaches to Fusion Energy

Table 4.1 Basic fusion reactions for technology ([43], p. 27, Figure 1.7).

Reaction	Q (MeV)	σ_{max} (barn)[a]	Energy peak (keV)
D + T → α + n	17.6	5	64
D + D → T + p	4.0	0.10	1250
D + D → ^{3}He + n	3.3	0.11	1750
D + D → α + γ	23.9		
D + ^{3}He → α + p	18.4	0.9	250
T + T → α + 2n	11.3	0.16	1000
P + ^{6}Li → α + ^{3}He	4.0	0.22	1500
P + ^{7}Li → 2α	17.4		
P + ^{11}B → 3α	8.7	1.2	550

a) 1 barn = 10^{-24} cm^2 = 10^{-28} m^2 = $(10 \text{ f})^2$.

Tokamak approach, we describe two small-scale demonstrations of deuterium fusion.

The first is based on field ionization of deuterium gas D_2 in a compact reaction chamber [45]. The resulting deuterons were accelerated to 80–115 keV and produced (small numbers) of fusion neutrons (the second reaction in Table 4.1) as they collided with the target, a layer of ErD$_2$, a solid with a large density of deuterons. This observation of fusion might suggest a reactor in which a solid surface of lithium or boron is bombarded with protons from a similar field ionization source. Such a direct solid-state fusion reactor was discussed in 1992 by Ruggiero [46], who concluded that useful amounts of power would not be available. The problem with this is that the probability of fusion is low because the D ions rapidly lose energy as they hit the Er ions, with their large numbers of orbiting electrons, in the target. Ionization of these electrons quickly slows the D ions to the point where fusion is impossible. In contrast, in the Tokamak DT plasma the ions have a very long mean free path.

The second demonstration of fusion on a small laboratory scale is based on creating "muonic" deuterium molecular D_2^+ ions, which can decay by D–D fusion and muon release [47], according to the 2D and 3D processes in Table 4.1. (A mu meson, muon, or μ is like an electron but 207 times more massive.) It is unlikely that either of these methods can evolve toward useful release of energy. At the moment, the two demonstrate only that controlled fusion, as monitored by release of neutrons, can be achieved under modest laboratory conditions.

4.1
Deuterium Fusion Demonstration Based on Field Ionization

In Chapter 1, we estimated that the potential energy of two protons in contact, $U = k_C e^2/r$, is about 0.6 MeV, but that protons having thermal energies in the lower range of tens of keV in the sun essentially accounted for its output of energy. As was explained in more detail at the end of Chapter 2, the particles are able to tunnel through the

(a)

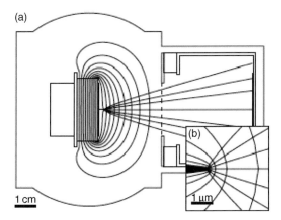

1 cm

(b) 1 μm

Figure 4.1 Sketch [45] of compact field-ionizing neutron generator operating in D_2 gas. The inset shows the 100 nm radius W tip which, when biased to 80 kV by the pyroelectric crystal, ionizes nearby deuterium gas to produce free deuterons. The free deuterons D^+ are accelerated into the ErD_2 target on the right, providing a high density of deuterons for the reaction.

Coulomb barrier, as explained by Gamow, from distances much larger than the contact spacing. The value for two deuterons is similar, but would formally be given from the nuclear radius formula $r = 1.07 fA^{1/3}$ with $A = 2$ and $f = 10^{-15}$ m. Therefore, the contact spacing is $2r_D = 2.14 f2^{1/3} = 2.7$ f, so that $U = k_C e^2/2r_D = 533$ keV.

Naranjo et al. [45] achieved D–D fusion in a small chamber filled with D_2 gas, by using a 100-nm-radius tungsten W tip biased to a positive voltage 80–110 kV. The electric field near the tip is large enough to break up the deuterium molecules and to ionize the resulting deuterium atoms, producing D^+ ions and electrons. The D ions are accelerated into an ErD_2 target and fusion neutrons were observed.

The apparatus is shown in Figure 4.1. A $LiTaO_3$ "pyroelectric" crystal (center left, in Figure 4.1) produces a strong electric field between the crystal and the ground plane. A pyroelectric crystal, as the temperature is varied, to give a large surface charge density, σ (in C/m^2), as electrons are pushed outward slightly from the crystal, and this means a large surface electric field $E = \sigma/2\varepsilon_0$. In this apparatus, the voltages appear with change in temperature of the pyroelectric crystal. Once ions are formed, they are carried into the target at energy 80 keV. The pyroelectric crystal, similar in nature to a ferroelectric crystal, is a clever way to achieve a high voltage in a small laboratory apparatus.

To generate ions, much larger localized ionizing electric fields E are generated near a tip of radius 100 nm (see inset). The electric field E at the surface of a metal sphere of radius a at voltage V is V/a, which in this case gives $E(a) = 80 \times 10^3/100 \times 10^{-9} = 0.8 \times 10^{12}$ V/m $= 0.8$ keV/nm at radius $a = 100$ nm. The field falls off as $E = Va/r^2$ for r larger than a.

The authors [45] indicate that $E = 25$ V/nm is sufficient for 100% field ionization, thus generating electrons and D^+ ions from D_2 gas. Thus, ionization rapidly occurs out a radius $R_{\text{ion}} = 100 \times [0.8 \times 10^3/25]^{1/2} = 0.565\,\mu$ from the tip, and the area of the ionizing hemisphere facing the target is $2\pi R_{\text{ion}}^2$. The rate of ion formation is the product of this area times $nv/4$, which is the rate, units $1/\text{m}^2\text{s}$, at which gas molecules cross unit area. Here, n is the density of deuterium molecules, of thermal velocity v, with another factor of 2 because each molecule releases two deuterons. The ion current I_i is then

$$I_i = 2\pi R_{\text{ion}}^2 (nv/4) 2e. \tag{4.1}$$

The ambient gas D_2 is dilute, such that the mean free path λ exceeds the dimension of the chamber, so that the D^+ ions fall down the potential gradient without collisions and impact the (deuterated) target at 80 keV energy. This energy is enough to drive the nuclear fusion reaction

$$D^+ + D^+ \rightarrow {}^3\text{He}^{2+}(820\,\text{keV}) + n(2.45\,\text{MeV}),$$

(where D is the deuteron, the nucleus of deuterium, ^{2}H).

In the compact device shown in Figure 4.1 [45], no external high voltages are needed. In this case, heating the LiTaO$_3$ crystal from 240 to 265 K, using a 2W heater, is stated to increase the surface charge density σ by $0.0037\,\text{Cm}^{-2}$. This should correspond to a surface electric field $E = \sigma/2\varepsilon_o = 3.7 \times 10^{-3}/(2 \times 8.85 \times 10^{-12}) = 0.209\,\text{V/nm}$. The authors state that in the device geometry (see Figure 4.1) this gives a potential of 100 kV. The observations, summarized in Figure 4.2, confirm D–D fusion, based on the observation of neutrons.

In Figure 4.2a, we see that a linear increase in temperature with time from 240 K to 280 K leads to crystal potential rising from zero to about 80 kV at $t = 230$ s. Figure 4.2b shows onset of the X-rays that come as electrons released from the ionization events fall onto the positively charged copper plate encasing the tantalate pyroelectric crystal. The X-ray energies, observed up to about 100 keV, can come only from a tip potential of the same order, which produces a local electric field sufficient to strip electrons from the deuterium gas. Figure 4.2c records the ionic current, presumably the sum of electron current into the copper and positive ion current into the right hand electrode, adding to 4 nA maximum. This number can be checked by elementary considerations involving the radius and surface area around the tip leading to certain ionization, the density of the gas at the stated pressure, and the number of deuterium molecules in random gas diffusion that would cross that surface per unit time, indicated in Equation 4.1. Finally, Figure 4.2d shows the measured number of neutrons per second. The satisfactory coincidence of the peaking of the several indicators at about 230 s makes it clear that fusion has occurred.

Let us look at these results from the simplified theoretical model developed in Chapter 2, leading to Equation 2.21. The tunneling transmission probability of the D through the Coulomb barrier (see Figure 2.3) at 40 keV is now estimated as

$$T = \exp(-2\gamma), \tag{2.21}$$

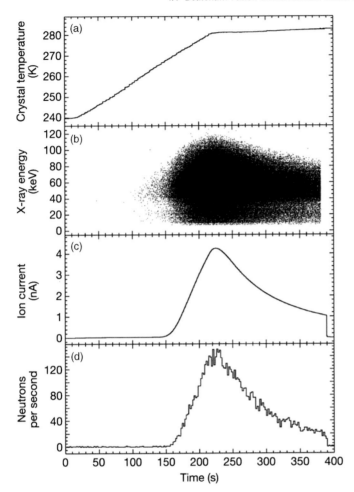

Figure 4.2 Documentation of one cycle of neutron generation in the compact device [45].

where, approximately,

$$\gamma = (2mE)^{1/2}/\hbar \left[(\pi/2)r_2 - 2(r_2 r_1)^{1/2} \right] \quad (2.24)$$

and r_2 is the classical turning point of the D. This is determined by the relation

$$40 \text{ keV} = k_C e^2 / r_2,$$

in the center of mass frame of the two deuterons, with accelerator voltage 80 kV, so that $r_2 = 36$ f. We have already seen that $r_1 = 2.7$ f. Using $m_R = 1.67 \times 10^{-27}$ kg, we find

$$\gamma = [(2mE)^{1/2}/\hbar] \left[(\pi/2)r_2 - 2(r_2 r_1)^{1/2} \right] = 1.62. \text{ Therefore } T_{\text{Gamow}}$$

$$= \exp(-2\gamma) = \exp(-3.239) = 0.0392.$$

The authors [45] measured 130 neutrons/s at peak, when the accelerating voltage was actually 115 kV, rather than 80 kV. To adjust the tunneling probability T_{Gamow} value to 115 keV, note (text near Equation 2.24) that the energy E dependence of γ is approximately $E^{-1/2}$. This gives tunneling probability $T_{Gamow} = \exp[-(E_G/E)^{1/2}]$.

For the D–D reaction in the compact chamber, this would imply $3.239 = (E_G/E)^{1/2}$ at $E = 40$ keV, so that $E_G = 419.6$ keV. At $(115/2)$ keV, then, $T_{Gamow} = 0.067$. (We neglect the small change in r_2.)

From our simplified approach in Chapter 2, the cross section for fusion is, for $r_2 = 36$ f, and $T_{Gamow} = 0.067$,

$$\sigma = \pi r_2^2 T_{Gamow} = 2.728 \times 10^{-28} \text{ m}^2 = 2.728 \text{ barn}. \tag{4.2}$$

This cross section value, for an energy near 100 keV, exceeds the maximum value quoted in Table 4.1, which is 0.11×10^{-28} m^2 at an energy of 1750 keV. We can attribute this discrepancy to a reaction probability $T = 0.04$, having to do with the details of the nuclear reaction, recalling that an analogous factor $T = 8.0 \times 10^{-24}$ was found for the p–p reaction on the sun.

From a more general point of view, it may be useful to look at the expected energy dependence of the cross section shown in Figure 4.3, in the right-most curve.

Figure 4.4 shows the dominant energy-dependent functions for nuclear reactions between charged particles in a thermalized plasma, as compared to a fixed energy beam as in the compact device of Figure 4.1. While both the energy distribution

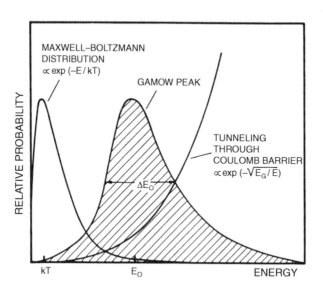

Figure 4.3 Right curve represents tunneling probability T_{Gamow} (Equation 2.21) as a function of center of mass energy. The compact fusion device provides D at a fixed energy ~40 keV, rather than a thermal Maxwell–Boltzmann distribution of energies as occurs on the sun or in a Tokamak reactor. The Gamow peak represents the optimal overlap of the energy distribution and the tunneling probability curves ([48], Figure 4.6, p. 159.

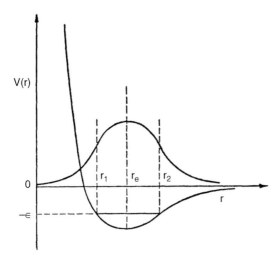

Figure 4.4 Ground-state potential V(r) and lowest vibrational wavefunction sketched for DDµ or DTµ ion. The large extent of the vibrational wavefunction suggests a chance for fusion, related to a nonzero probability of nuclear spacing near zero [47].

function (Maxwell–Boltzmann) and the quantum mechanical tunneling function through the Coulomb barrier are small for the overlap region, the convolution of the two functions results in a peak (the Gamow peak) near the energy E_o, giving a sufficient probability to allow a significant number of reactions to occur. The energy of the Gamow peak is generally larger than $k_B T$. In the compact device of Figure 4.1, the energy is fixed by the accelerating potential, so the thermal distribution of energies in Figure 4.4 is replaced by a single energy.

Returning to the results (Figure 4.3) from the compact fusion device, we can find the experimental probability P_f per incoming D of fusion. The result in the experiment, since the measured 4.4 nA corresponds to a deuteron flux of $4.4 \times 10^{-9}/(1.6 \times 10^{-19}) = 2.75 \times 10^{10}$ D/s, is

$$P_f = 130/(2.75 \times 10^{10}) = 4.72 \times 10^{-9} \qquad (4.3)$$

at accelerator voltage 115 kV.

We can estimate the density of deuterons in the ErD$_2$ target as $n_D = 6 \times 10^{28}$ m^{-3}. The mean free path for fusion then is $\Lambda = 1/n_D\sigma = 0.061$ m. However, the deuterons rapidly get slowed down, and do not in fact penetrate more than a few atomic layers. The energy loss $dE/dx = -\alpha$ that comes from electron excitation as the energetic incoming ion proceeds into the target is on the order of 200 MeV/mm. We can estimate the distance of penetration as $x \sim 0.5\, E_{in}/\alpha$. Taking $E_{in} = 115$ keV, we find $x = 2.8 \times 10^{-7}$. If fusion occurs certainly for 0.61 m, then in distance x the chance of fusion is

$$P_f(x) = x/0.061 = 4.7 \times 10^{-6}. \qquad (4.3a)$$

This value is larger than observed. The predicted efficiency from an energy release point of view is $(3.5 \text{ MeV}/115 \text{ keV}) \times 4.7 \times 10^{-6} = 1.43 \times 10^{-4}$. These numbers are approximately applicable to the deuteron experiment [45], but they predict a higher yield than was observed. From these results, confirmed by Naranjo et al. [45], there seems not much hope for making a practical fusion reactor from an energetic beam hitting a solid or liquid target, because of the rapid energy loss, as was earlier predicted by Ruggiero [46].

4.1.1
Electric Field Ionization of Deuterium (Hydrogen)

Ionizing an atom, as in the compact apparatus of Figure 4.1, is done with an electric field, and the question is how large a field is required? The same question arises in a Tokamak device where it is necessary to ionize the deuterium gas to form a plasma. So the rate of "field ionization," which is governed by an electron tunneling process, is important.

The electronic properties of deuterium are the same as hydrogen H, which differs only by the additional (uncharged) neutron mass in the nucleus. Thus, the simplest estimate of the required electric field E to ionize deuterium might be the field of the proton at the Bohr radius in H, which is $E = k_C e/a_o^2 = 514.5 \text{ V/nm}$.

This is an overestimate: a much smaller electric field, $E \approx 25 \text{ V/nm}$, quickly ($\sim 10^{-15}$ s) removes the electron by a process of tunneling.

In an electric field E, the potential energy of the electron has a term $U(x) = -eEx$, so that at a spacing $x^* = E_o/eE$, the electron will have the same energy as in its ground state, and can tunnel out. Here E_o is the electronic binding energy, 13.6 eV for deuterium, as well as for hydrogen. How quickly will this happen?

The electron can tunnel through the barrier that extends from $x = a_o$ to x^*. In detail this is a difficult problem to solve, but a simplified treatment is possible. The earliest estimate of the lifetime of H against field ionization was given by Oppenheimer in 1928. Oppenheimer's notable estimate [49] is that for H in a field of 1000 V/m the lifetime τ is $(10^{10})^{10}$ s.

We can use a simplified, one-dimensional, model to estimate the E-field ionization rate of H. We will find that for $E = 25 \text{ V/nm}$, the field ionization lifetime of hydrogen and deuterium is on the order of 10^{-15} s, while, at $E = 2.5 \text{ V/nm}$, τ is extremely long, estimated as 6×10^{33} s or 1.9×10^{26} y.

The ground state with $E = 0$ is spherically symmetric, depending only on r. With an electric field $E_x = E$, approximate the situation as depending only on x: electron potential energy

$$U(x) = -k_C e^2/x - eEx, \qquad (4.4)$$

where $k_C = 9 \times 10^9 \text{ Nm}^2/\text{C}^2$, with E in V/m. Assume that the ground-state energy is unchanged by the electric field, and remains $-E_o = -13.6 \text{ eV}$.

The electron barrier potential energy is

$$V_B(x) = U(x) - E_o. \qquad (4.4a)$$

For electric field $E=0$, the electron is assumed to have energy $U=-E_o$ at $x=a_o=0.053$ nm, the Bohr radius. When $E>0$, there is a second location, near $x_2=E_o/eE$, where U is again E_o. The problem is to find the time τ to tunnel from $x=a_o$ to $x=x_2=E_o/eE$ through the potential barrier $U(x) - E_o$. For small field E, the maximum height of the barrier seen by the electron is $\approx E_o = 13.6$ eV, and the width t of the tunnel barrier is essentially $\Delta x = t = x_2 = E_o/eE$.

If the barrier potential $U(x)$ were V_B (a square barrier independent of x), the tunneling transmission probability T would be

$$T = \exp[-2(2mV_B)^{1/2}\Delta x/\hbar] \tag{4.5}$$

for barrier width Δx and $\hbar = h/2\pi$, with h Planck's constant.

A more accurate approach would replace $(V_B)^{1/2}\Delta x$ by $\int [U(x)-E_o]^{1/2}$ dx, this is known as the WKB approximation, and was also used in the Gamow tunneling probability, see Equation 2.22.

We will simplify further, and calculate on the basis of the *average value* of the barrier: approximate the barrier function $U(x) - E_o$ by half its maximum value, $\frac{1}{2} V_{BMAX}$, over the width Δx.

The escape rate, $f_{escape} = 1/\tau$, with τ the desired ionization lifetime, is $Tf_{approach}$. The orbital frequency of the electron will be taken from Bohr's rule for the angular momentum $L = mvr = h/2\pi$, so $f_{approach} = h/[(2\pi)^2 ma_o^2] = 6.5 \times 10^{15}$ s^{-1}. The working formula $f_{escape} = 1/\tau$ is

$$f_{escape} = f_{approach} \exp[-2(mV_{BMAX})^{1/2}\Delta x/\hbar]. \tag{4.6}$$

To find Δx one solves the quadratic $-E_o = -ke^2/x - eEx$ (E is the electric field):

$$\Delta x = [(E_o/e)^2 - 4k_C eE]^{1/2}/E. \tag{4.7}$$

(This limits the applicability of the approximation to $E < 32.1$ V/nm, where $\Delta x = 0$.)

So we have the width of the barrier, and now we want to find the location of its peak and the peak value.

The equation for U is $U(x) = -k_C e^2 x^{-1} - eEx$. Taking the derivative with respect to x, dU/d$x = U'$, we set $U' = 0$ to find the barrier peak. Thus,

$$U' = k_C e^2 x^{-2} - eE = 0 \text{ at}$$

$$x' = (k_C e/E)^{1/2}, \tag{4.8}$$

which locates the peak of the barrier. Plugging this value, x', into the expression for U, we find the peak value of U, U_{max}, to be

$$U_{max} = -k_C e^2 (E/k_C e)^{1/2} - eE(k_C e/E)^{1/2} = -2e(k_C Ee)^{1/2}. \tag{4.9}$$

Thus, the peak height of the barrier is

$$V_{BMAX} = E_o - 2e(k_C eE)^{1/2}. \quad (4.9a)$$

Next, we evaluate f_{escape} for $E = 25$ V/nm, using

$$f_{escape} = f_{approach} \exp[-2(2mV_{BMAX}/2)^{1/2}\Delta x/\hbar], \quad (4.6)$$

taking the average barrier as $V_{BMAX}/2$. To evaluate the square bracket term one finds, at 25 V/nm,

$$\left[-2(mV_{BMAX}/2)^{1/2}\Delta x/\hbar\right] = -2.35. \quad (4.10)$$

The escape frequency is then
$f_{escape} = 6.5 \times 10^{15} \exp(-2.35) \text{ s}^{-1} = 6.2 \times 10^{14} \text{ s}^{-1}$, and the ionization time is

$$\tau = 1.61 \times 10^{-15} \text{ s}, \quad \text{for } E = 25 \text{ V/nm}. \quad (4.11)$$

If we repeat the analysis of f_{escape} from the working formula for $E = 2.5$ V/nm, we find that the square bracket is -118.8. The lifetime is about

$$\tau = 6 \times 10^{33} \text{ s} \quad (\text{for } E = 2.5 \text{ V/nm}) \quad (4.12)$$

or about 1.9×10^{26} y. The simple tunneling model predicts a sharp cutoff in the ionization rate for electric fields falling from 25 V/nm to 2.5 V/nm, as observed.

Tunneling rates are sensitively dependent on the parameters related to the barrier. This is observed both experimentally and in model calculations like the one we have given.

4.2
Deuterium Fusion Demonstration Based on Muonic Hydrogen

This section is based on Equations 3.36–3.42 in Chapter 3. The hydrogen molecule ion has the same defining equations as the molecular ion based on two deuterons and one muon, which we now describe.

The muon has charge $-e$ and mass 207 m_e. Although its lifetime is only 2.2 µs, the properties of the corresponding "muonic" hydrogen and deuterium atoms are well known. The reduced mass $m_r = mM/(m + M)$ enters the Bohr radius in the denominator, and therefore multiplies the Bohr energy. Focusing on the "Dµ" formed by a muon and a deuteron D, the reduced mass is $m_r = 207 \times 2 \times 1836/[207 + 2(1836)] = 196 \, m_e$. The equivalent Bohr radius is 270 f and the binding energy is 196×13.6 eV $= 2.666$ keV. The Dµ atom is smaller by a factor 196 than hydrogen and has a binding energy 196 times larger. If a beam of muons passes through a gas of hydrogen atoms or hydrogen molecules, the electrons will be ejected with 2.66 keV energy and the muonic atoms will form. In dense hydrogen exposed in this way, it is observed that muonic DDµ$^+$ molecule ions form. For example,

$$\mu + DeeD \rightarrow DD\mu + 2e.$$

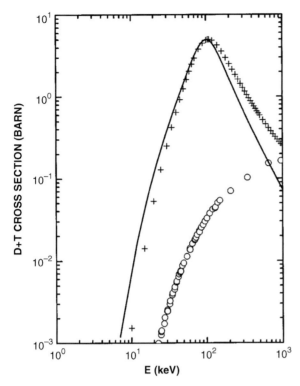

Figure 4.5 Measured cross section [50] for D–T reaction (solid curve) compared to resonant tunneling model (+) and measured values for D–D reaction (circles).

It is experimentally known that the lifetime of the DDμ for fusion of the D particles is about 1.5 ns, based on detecting the neutrons that are emitted by the fusion. It is also known that during the 2200 ns lifetime of a single muon in a dense deuterium gas several hundred cycles of molecule formation and fusion can occur!

The DDμ ion can be understood by scaling the properties of the HHe (H_2^+) ion as described in Equations 3.36–3.39. The scaling of the Bohr radius and Bohr energy leads to an equilibrium spacing $R = r_e = 2.4 a_o/196 = 648$ f. The binding energy of the scaled ion is 519.4 eV. (We will see later in this chapter that this binding energy is reduced by about half by the positive confinement energy of the vibrational motion.)

Following the discussion of Equation 3.38a, we equate the energy E with the potential sketched in Figure 4.5. First estimate the vibrational frequency $\omega = (E''/m_{red})^{1/2}$, where the curvature $E'' = d^2E/dr^2$. We scale the curvature quantity

$$d^2 E(r_e)/dr^2 = E''(r_e) = 0.1257\, E_o/a^2 \tag{4.13}$$

with Bohr energy and Bohr radius to get the DDμ oscillation frequency

$$\omega = (E''/m_{red})^{1/2} = (0.1257\ 13.6\ 196\ 1.6\ 10^{-19}/1.67\ 10^{-27})^{1/2}/270 = 6.64 \times 10^{17}\ \text{s}^{-1}. \tag{4.14}$$

This high frequency corresponds to a "zero point" energy

$$\hbar\omega/2 = 217.8 \text{ eV}, \qquad (4.15)$$

roughly half the electronic binding energy, as suggested in Figure 4.5. (The zero point energy is a confinement energy, not present in the classical oscillator, that is implied by the uncertainty principle $\Delta p \Delta x \geq h/2\pi$ as applied to the momentum of the oscillator.) The corresponding spring constant

$$K = m\omega^2 = 0.735 \times 10^9 \text{ N/m}. \qquad (4.16)$$

In Figure 4.5, it appears that $V(r)$ is zero near D–D spacing $r_e/2 = 324\,\text{f}$. If we assume a parabolic variation of V in the range $r_e/2 < r < r_e$, we can make an alternative rough estimate

$$K = 1.58 \times 10^9 \text{ N/m}, \qquad (4.16a)$$

which is not too different from (4.16), considering the approximate nature of the analysis.

We can now estimate the time for fusion T_f, using the properties of this DDµ quantum mechanical oscillator.

The formal problem of the 1D oscillator in Schrodinger's equation gives

$$E_n = (n + \tfrac{1}{2})\hbar\omega, \qquad (4.17)$$

where $\omega = (K/m)^{1/2}$ as above and $n = 0, 1, 2, \ldots$

We can find the zero point vibrational width (see Figure 4.5), dimension in meters,

$$\sigma = (\hbar/m\omega)^{1/2}, \qquad (4.18)$$

which enters the probability distribution for the $n=0$ (zero point) vibration,

$$P_{n=0}(x) = \sigma^{-1}(1/\pi)^{1/2} \exp-(x/\sigma)^2, \qquad (4.19)$$

where $x = r - r_e$. We can evaluate the vibrational width of motion

$$\sigma = (\hbar/m\omega)^{1/2} = [6.6 \times 10^{-34}/(2\pi\, 1.67\, 10^{-27} 6.64 \times 10^{17})]^{1/2} = 307.8\,\text{f}. \qquad (4.18a)$$

At this spacing, $r = r_e - \sigma = 340.2\,\text{f}$, in the ground-state motion of the oscillating masses, the probability is reduced to $1/e = 0.368$. This width is large, around half the equilibrium spacing, consistent with the observed DDµ decay by fusion of the deuteron particles.

As a first rough estimate of the fusion rate, we can simply evaluate the probability $P'(0 - r_1)$ that the oscillator spacing is in the range $0 < r < r_1$ ($r_1 \approx 2.7\,\text{f}$, corresponding to contact between the two deuterons), and multiply by the oscillator frequency.

4.2 Deuterium Fusion Demonstration Based on Muonic Hydrogen

However, since this assumes a parabolic potential rather than Coulomb repulsive potential, it will likely overestimate the fusion rate. The probability of D–D spacings in the range 0–r_1 is the integral of $P(x)$ over that range, which can be approximated as

$$P'(0-r_1) \approx r_1 \sigma^{-1}(1/\pi)^{1/2} \exp-[(r_e-r_1)/\sigma]^2 \tag{4.20}$$

$= (2.7/307.8)0.564 \exp-[(648-2.7)/307.8)^2$. The result is $P'(0-r_1)$
$= 8.77 \times 10^{-3} \quad 0.564 \quad 0.0123 = 6.08 \times 10^{-5}$.

The rate of fusion is then estimated as $R = (\omega/2\pi) \, P'(0-r_1) = 6.43 \times 10^{12} \, s^{-1}$, corresponding to a lifetime

$$T_f = 1.55 \times 10^{-13} \, s. \tag{4.21}$$

This is shorter than the observed lifetime, which is 1.5 ns.

The correct solution is difficult, but was described by Jackson [47]. While the full solution is more advanced than we can present here, we suggest an approximation that improves on what we have just done.

The improved approximation is to replace the harmonic oscillator potential with the direct Coulomb potential, for the range between the classical turning point $R = r_e - \sigma = 340.2$ f and the contact radius $R = r_1$. (In physical terms, this means that we neglect the muon charge density in that region, removing screening of the Coulomb potential.)

The corresponding Gamow transmission factor

$T_{Gamow} = \exp-2\gamma$, where, for $r_2 = 340.2$ f, $E = 4.233$ keV, $r_1 = 2.7$ f,
$$\tag{4.22}$$

$$\gamma = (2mE)^{1/2}/\hbar[(\pi/2)r_2 - 2(r_2 r_1)^{1/2}] = 6.78$$

and $T_{Gamow} = 1.28 \times 10^{-6}$. In this case, the integration of the oscillator probability function is from 0 to $r_2 = 340.2$ f and is about $P' = 0.023$. The resulting rate of fusion $R = (\omega/2\pi) \, P'(0-r_2) \, T_{Gamow} = 3.1 \times 10^9$, so

$$T_f = 0.32 \times 10^{-9} \, s. \tag{4.23}$$

This is closer, but still below the observed value, which is 1.5 ns ([43], p. 26). We attribute the difference to a reaction probability factor $T = 0.21$, which we can interpret as the fraction of the collisions of the two deuterons at the contact point that lead to fusion, an effect of the nuclear physics of the two particles. The estimate here is of the same sense but numerically somewhat larger than the value $T = 0.04$ we mentioned earlier.

The influence of the details of the nuclear reaction, suggested here, is supported by the well-known large difference between this lifetime and the fusion lifetime of the similar DTμ ion, $T_f = 7 \times 10^{-13}$ s. This represents an increase in the reaction rate R by a factor 21. (This is similar to the increase in the DT above the DD cross section in

Table 4.1.) The Gamow factors are almost the same for the DD and the DT cases, so we need to learn more about the nuclear reactions.

From the point of view of our calculation, adapting the molecular properties from the DD case to the DT case will change in our estimate of T_f only slightly. This can be seen from the larger reduced mass entering the calculation of γ. For DT, the reduced mass is $m_r = (2 \times 3)/(2 + 3) = 1.2$, a 20% increase, which will increase γ and reduce the tunneling rate. The parameter r_1, however, is increased from 2.7 f to 2.89 f, leading again to a small change, but in the opposite sense to that of the effective mass change. The resulting changes will be small, suggesting that details of the nuclear tunneling physics, not built into our estimate, may be important.

Why is the D–T fusion reaction easier to initiate than the D–D reaction? The difference is that the DT system has a "resonant state" at energy 114 keV, as shown in data [50] of Figure 4.6.

The solid curve in Figure 4.6 is the measured cross section for the D–T reaction. Not only is the cross section much larger than the data for the D–D reaction (circles) but it also has a clear peak, at a resonant energy 114 keV. The physical situation is as described by Figure 2.3, imagining a D coming to collide with a T triton particle, except that there is a well-defined "resonant state" at positive energy $E = 114$ keV. This is a metastable state, which can decay either by back reflection of the D, or by forming the ^{4}He + n final state. This final energy is negative (bound and stable), analogous to the -2.2 MeV marked in Figure 2.3 denoting the D ground state.

There are, in effect, two tunnel barriers in this problem. One is the Coulomb barrier shown in Figure 2.3 and the second effective barrier is between the resonant

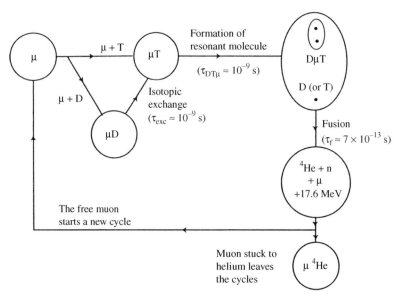

Figure 4.6 μ-catalysis cycle for the DT gas mixture, as irradiated with mu mesons. Mean reaction times are indicated for each process [43].

state and the final state. The large increase in cross section, approximately the inverse of the Gamow factor $T_{Gamow} = 0.067$ found for 57.5 keV in connection with Equation 4.2, namely, 14.9, is expected for resonant tunneling, *when the incoming particle energy matches that of the resonant state.*

In the physics analogy, one has two barriers separated by a potential well. The interesting effect comes when the well has a bound state at the energy of the input tunneling particle. In this case, the transmission probability (across the whole system) becomes 1, which corresponds to setting the Gamow factor to 1 as we mentioned. There is thus a strong peak in the transmission probability $T(E)$ at the resonant energy $E = E_0$. In more detail, near the energy of the resonant state E_0

$$T(E) = \Gamma^2/[\Gamma^2 + (E-E_0)^2]. \tag{4.24}$$

The (+) symbols shown in Figure 4.6 as calculated by Li et al. [50] are in good agreement with the measurements and establish the resonance state interpretation of the large enhancement of the DT cross section over the DD cross section peaking at 114 keV. In Equation 4.24, the energy width is of the form $\Gamma = \hbar/\tau$ where τ is the lifetime of the resonant state. The lifetime is the inverse of the rate of tunneling out of the state, $\omega = \omega_{attack} T_{Gamow}$, so for $\Gamma^2 \ll (E-E_0)^2$ we get

$$T(E) \approx (\hbar \, \omega_{attack} T_{Gamow})^2/E^2. \tag{4.24a}$$

This is not a familiar formula, but does indicate that the transmission factor doubly enters the rate as $(T_{Gamow})^2$ away from the resonance, when the barrier width is effectively doubled.

4.2.1
Catalysis of DD Fusion by Mu Mesons

Our discussion leaves no doubt that molecular ions of DDμ and DTμ will be strongly bound until destroyed by fusion or by the decay of the mu meson, after 2200 ns. Fusion occurs in these ions on timescales about 1.5 ns for DD and 0.7 ps, for DT. In both cases the muon is released, following a fusion event, and, after some delay, forms a new DDμ or DTμ ion. The role of the muon is as a catalyst.

The detailed cyclic process ([43], p. 26) for DT is shown in Figure 4.7. A dense gas mixture of D and T is bombarded with muons, momentarily forming Dμ and Tμ atoms.

These atoms form molecules and molecular ions, as shown in Figure 4.7. The DTμ molecule formed in a time about 1 ns, and promptly undergoes fusion in 0.7 ps. After fusion, the muon is released, and again available to catalyze further fusions. The cycle described takes place in about 5 ns, so in the muon lifetime about 440 fusion events could occur. However, as suggested in Figure 4.7, a small fraction, about 0.006, of the fusion muons are not released, but remain attached to the helium fusion product. On this basis, the number of fusions per muon is estimated as 120. Experiments have shown up to 200 fusion events per muon.

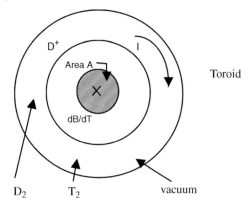

Figure 4.7 Essential elements of Tokamak reactor, top view. A toroidal vacuum chamber is filled with deuterium and tritium gas at about $10^{20}/m^3$ density. Heating is applied to form a neutral high-conductivity plasma of deuterons, tritons, and electrons. A toroidal magnetic field B, not shown, is in the range 1–5 Tesla, produced by current through a superconducting coil wound around the toroid. The plasma current I is induced by a changing magnetic flux, AdB/dt, through the hole of the donut. The toroidal device can be regarded as a solenoidal coil of radius b, whose length is $2\pi a$, closed on itself to form a torus of volume $V = 2\pi a \pi b^2$.

This catalytic approach to fusion has been measured and modeled, but has not produced more fusion energy than needed to initiate the process.

In summary, muon-catalyzed fusion (μCF) is a process allowing *nuclear fusion* to take place at *temperatures* significantly lower than the temperatures required for *thermonuclear fusion*, even at *room temperature* or lower. Although it can be produced reliably with the right equipment and has been much studied, it is believed that the poor energy balance will prevent it from ever becoming a practical power source. However, if *muons* (μ^-) could be produced more efficiently, or if they could be used as *catalysts* more efficiently, the *energy balance* might improve enough for muon-catalyzed fusion to become a practical power source.

The observed time for fusion in DDμ is 1.5 ns, and shorter in DT μ, while the lifetime of μ is 2200 ns. This would allow for 1400 cycles, emitting one neutron plus 3 MeV per cycle, but the best seen is only 200. This difference could make the process useful for energy production. At present, it seems that something else is limiting the neutron emission rate, maybe the time for a new DDμ or DTμ to form from the released μ, in the sample mixture of deuterium, DeeD and related DeeT molecules. This is perhaps limited by diffusion.

4.3
Deuterium Fusion Demonstration in Larger Scale Plasma Reactors

The essential elements of a Tokamak reactor are sketched in Figure 4.8. Here, a toroidal vacuum chamber contains positive ions and electrons. A solenoid runs

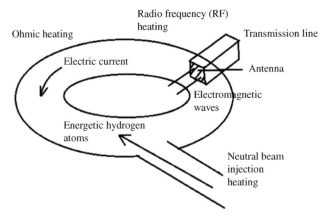

Figure 4.8 Sketch indicating radio frequency (microwave) and neutral beam injection heating, in addition to ohmic heating of gas mixture to reach the plasma operating temperature. Assuming the toroidal magnetic field 5 T, the electron cyclotron resonance frequency was found in the text to be 140 GHz (http://www-fusion-magnetique.cea.fr/gb/fusion/principes/principes05.htm).

through the hole of the "donut" and its magnetic field is ramped to produce an EMF round the loop.

4.3.1
Electrical Heating of the Plasma

Conventional electric plasma heating $P = I^2 R$ is provided by Faraday's law of induction of heating current from changing magnetic flux $d/dt\ (BA) = -\varepsilon = IR$. Fusion occurs at temperatures near 150 million K. When fusion occurs, released energy is collected from the energetic neutron emissions, in a surrounding blanket and heat exchanger (not shown, see below), while the charged reaction products are retained in the plasma by the magnetic confinement, and further heat the plasma.

We will model a simplified Tokamak, assuming a pure deuteron–electron plasma at density $n = 10^{20}$, $T = 150 \times 10^6$ K (10 times higher than the sun's core).

Consider a torus of volume 489.5 m^3 (about 500) and this has major radius 6.2 m and minor radius 2 m. Think of this as a cylinder of length 2π 6.2 m and cross-sectional area $\pi\ 2^2$. Using the formula for the resistance R of a length L with cross section A, $R = \varrho L/A$, the resistance around this torus, containing the plasma as described, is $R = \varrho\ 2\pi\ 6.2/\pi\ 2^2 = 3.1\ \varrho$ ohms.

We have described in Chapter 2 the power density from fusion in the sun, at some length. Our approach here will be to scale the analysis from the sun to the situation of the model Tokamak plasma. Scaling is an attractive approach with the assumption that a successful analysis allows extrapolation to a similar situation with modified parameters.

4.3.2
Scaling the Fusion Power Density from that in the Sun

In Chapter 2, the power output from the sun is analyzed from the point of view of proton–proton fusion, starting with Equation 2.21, and Equations 2.25–2.28, then finally

$$P = 0.5\, N_p^2 \langle v\sigma \rangle Q, \tag{2.33}$$

for the power per cubic meter generated, 313 W/m^3 in the core of the sun.

Our program here is to scale this number to find the power density in a D–D reaction Tokamak reactor, assuming deuterium density $N_D = 10^{20}$ m^{-3} (see Figure 3.18), to be compared with the sun's proton density $N_p = 3.11 \times 10^{31}$ m^{-3}. This, along with the change in the reaction probability factor T from 8×10^{-24} (Equation 2.31) to $T = 0.1$. Assume a factor of 10 increase in T from 15 million K to 150 million K. Scaling must include the r_2 parameter, noting that the mass of the deuteron is twice that of the proton, and the temperature enters the velocity and also the cross section for the geometric collision. Finally, if the Tokamak has a volume 500 m^3, how much power does it release in fusion reactions (assume $Q = 3.5$ MeV for the DD reaction)?

The scaling ratio, P'/P, works out to be

$$P'/P = (10^{20}/3.11 \times 10^{31})^2 (1/10^{3/2})(T'_{\text{Gamow}}/10^{-8})(0.1/8 \times 10^{-24})$$
$$(3.5\,\text{MeV}/26.2\,\text{MeV}) = 54.01,$$

where $T'_{\text{Gamow}} = 9.9 \times 10^{-4}$. (The second factor here comes explicitly from the change in temperature entering v and σ.) The value for T'_{Gamow} follows Equation 2.24 with parameters $r_1 = 3$ f, $r_2 = 111.3$ f, $E = 12.93$ keV, and $m_r = 1 \times m_p$.

So the predicted Tokamak power density, assuming pure deuterium fuel, is 16.9 kW/m^3 and its power output is 8.45 MW.

4.3.3
Adapt DD Plasma Analysis to DT Plasma as in ITER

To convert this result to a DT plasma, approximately, we consider two aspects. First, is the factor 15 increase in the DT cross section over the DD cross section, and, second, the larger Q energy release, 17.6 MeV, versus 3.5 MeV. Thus, if the plasma were a DT plasma, the projected Tokamak power output would be increased by 75, to

$$Q = 634\,\text{MW}. \quad (\text{DT plasma at 150 million K, 500 m}^3). \tag{4.25}$$

This estimate is close to those projected for the ITER Tokamak, which has a larger volume, 880 m^3, by virtue of its larger, heart-shaped, rather than circular, cross section.

Let us continue to calculate properties of the assumed Tokamak, starting with the energy U needed to heat the 500 m^3 deuterium to 1.5×10^8 K, where we know it is 12.93 keV per particle. In thermal equilibrium, the plasma contains equal numbers of ions (deuterons plus tritons) as electrons, and each has average energy 12.93 keV. Thus,

$$U = 2(N_D + N_T)k_B T \tag{4.26}$$

$= 2 \times 10^{20} \times 500 \times 12.93 \times 10^3 \times 1.6 \times 10^{-19}$ J $= 206.8$ MJ $= 57.5$ kWh. The market cost of this energy, at 14c/kWh, the price of electricity in New York City, is $8.05. This is small, and the key is the assumption that only the gas is heated, the containment (to be described) allows the walls and the whole containment vessel to remain cool.

We might next ask how long can the reactor run at 634 MW before running low on fuel? Arbitrarily, let us find $T_{1/2}$, when half the energy of the fusion reactions is consumed, assuming the D–T reaction generating 634 MW:

$$P\,T_{1/2} = 0.5 \times 10^{20} \times 500 \times 17.6 \times 10^6 \times 1.6 \times 10^{-19} \text{ J} = 7.04 \times 10^{10} \text{ J, so}$$

$T_{1/2} = 7.04 \times 10^{10}$ J$/634 \times 10^6 = 111$ s, not quite 2 min.

So the reactor will definitely need a continuous feed of DT mixture. (In the planned ITER device, the T (triton) generation is accomplished by feeding Li into system, which generates T (tritium) by absorption of neutrons.)

We should confront an issue of how the fusion power is harvested. While the reaction of DT produces 17.6 MeV, this energy is shared between a neutron and an alpha particle (first line in Table 4.1). In the frame of the fusion event, the momentum of the particles will add to zero, so $4V = -v$, with V and v, respectively, the speeds of the alpha (mass 4) and the neutron (mass 1). The kinetic energy of the neutron is $1/2\, mv^2$ and that of the alpha is $1/2\,(4m)(v/4)^2$. On this basis, the neutron gets 0.8 of the kinetic energy, which it carries out of the reaction zone into the containment structure. We will mention next how the alpha particle is trapped into circling the magnetic field lines, and thus returns its heat energy back into the plasma. The ^{4}He is a waste product, termed "ash" in the literature of Tokamaks. The neutrons are equally emitted in all spherical directions, and their motion is not at all impeded by the magnetic field.

Another aspect to be confronted is that we assume a uniform temperature distribution in the full torus. The major and minor radii a, b and volume should realistically be interpreted as the dimensions of the hot part of the plasma. Indeed, it is essential to establish a large radial temperature gradient in the cross section of the torus to keep the walls cool, except for the 14.1 MeV fusion neutrons that carry the fusion energy out of the reactor.

Another energy cost is the toroidal magnetic field; we assume it is 5 Tesla uniformly in the cross section πb^2. The energy density of the 5 T magnetic field B,

$$u_B = B^2/2\mu_0 \tag{4.26a}$$

$= 9.94 \times 10^6$ J/m^3, where $\mu_0 = 4\pi \times 10^{-7}$ N/A^2. Therefore, the energy of the whole uniform magnetic field is $U_B = 4.97$ GJ, with cost at $0.14/kWh of $193.42. The pressure of the magnetic field on the walls is $u_B = 9.94 \times 10^6$ J/m$^3 = 9.94 \times 10^6$ N/m^2.

Since a Pascal is 1 N/m², and one atmosphere (1 bar) is 101 kPa, the pressure is 98.4 atmospheres. So the magnetic vessel has to be very strong indeed.

The production of the field can be understood from the formula for the B field in a long solenoid, $B = \mu_0 \, nI$, where n is the number of turns of wire per meter and I is current. Once established, the field has no further energy cost because the current is maintained in a superconducting magnet.

The result of D–T fusion as we have seen is an alpha particle of energy $K = 0.2 \times 17.6 \text{ MeV} = 3.52 \text{ MeV} = {}^1/_2 \, 4m_p V^2$. The velocity of the alpha is thus $V = (2K/4m_p)^{1/2} = 1.3 \times 10^7$ m/s. The basic force

$$\mathbf{F} = q\mathbf{V} \times \mathbf{B} = m\mathbf{a} \tag{4.27}$$

with m is the mass. The acceleration

$$a = V_t^2/R \tag{4.28}$$

for a circular orbit of radius R, with V_t the component of the velocity transverse to the B field, gives us

$$R = mV_t/qB = (2\,mK)^{1/2}/qB \tag{4.28a}$$

for the radius R of the circular orbit, at kinetic energy K, around the B field lines. The path of the charged particle is a spiral, whose orbital radius depends on the angle of the velocity vector relative to the magnetic field B. A particle moving perpendicular to the field has the largest orbital radius. The frequency of the orbital motion is

$$\omega_L = qB/m, \tag{4.29}$$

which is called the Larmor frequency. For the alpha particle $^4\text{He}^{2+}$ released in the DT reaction at 5 Tesla, we find for the maximum radius,

$$R = (4 \times 1.67 \times 10^{-27} \times 1.3 \times 10^7)/(2 \times 1.6 \times 10^{-19} \times 5) = 0.054 \text{ m}. \tag{4.30}$$

This means that the alpha particle, charge $2e$ and mass $4m_p$, created in the plasma will at most move 0.054 m toward the wall of the toroidal container. Only those alpha particles created within 5.4 cm of the wall, and moving specifically in the radial direction, will have a chance of colliding with the wall. Since the emission directions of the alphas are randomly distributed, an even smaller probability of wall collisions will occur. This is the basis of the confinement.

The Larmor frequency for the electrons is of interest in regard to heating the plasma. It is

$$\omega_L = qB/m = 1.6 \times 10^{-19} \times 5/9.1 \times 10^{-31} \text{ rad/s}$$
$$= 8.79 \times 10^{11} \text{ rad/s} = 140 \text{ GHz}, \tag{4.31}$$

corresponding to radiation wavelength $\lambda = c/f = 2.14$ mm. This corresponds to F- or D-band microwave radiation.

Electrons are in thermal equilibrium with $K=12.93\,\text{keV}$ on average. Thus, $R=(2mK)^{1/2}/qB=76.7\,\mu\text{m}$, and we see that thermal electrons are controlled even more tightly, to a smaller radius, and move closely along the toroidally oriented magnetic field lines. So, the charged particle system, closed on itself, is like an infinite 1D system. The electron thermal velocity v in the plasma is of interest, and is $v=(k_BT/m_e)^{1/2}=4.77\times 10^7\,\text{m/s}$, about 0.16 c. The motion of these electrons is not much affected by the theory of special relativity, because the relevant factor $[1-(v/c)^2]^{-1/2}=1.012$ is close to 1, indicating Newton's laws govern the motion.

We want to learn about the electrical conductivity of the plasma, which is limited by the mean free path for scattering of the electrons by the ions. Assume the same parameter r_2 as was used in Chapter 2 (111.3 f at 12930 eV) determines the collision cross section. So for the electron mean free path, we find

$$\lambda_e = 1/[N_D\pi(r_2)^2] = 257\,\text{km (for electrons)}. \tag{4.32}$$

This is an amazingly long mean free path, indicating that the electrons orbit around the torus 6600 times between collisions. (We will see below that there is a correction that reduces this by a factor 21.1.) The mean time between collisions, $\tau=\lambda_e/v=5.38\,\text{ms}$. Now, we can find the mobility

$$\mu = e\tau/m = 9.47\times 10^8\,\text{m}^2/\text{V s} \tag{4.33}$$

and the electrical conductivity

$$\sigma = ne\mu = 1.52\times 10^{10}. \tag{4.34}$$

The resistivity of the plasma at $1.5\times 10^8\,\text{K}$ and $n=10^{20}$ deuterons/m^3 is thus

$$\varrho = 1/\sigma = 6.6\times 10^{-11}\,\Omega\text{m}. \tag{4.35}$$

We can neglect the contribution of the ions to the conductivity because their mobility is at least 1835 times smaller due to the mass increase between proton and electron.

We can now find the resistance R for the torus plasma at its operating point as $R=\varrho 2\pi\,6.2/\pi 2^2 = 3.1\,\varrho$ ohms,

$$R = 0.204\,\text{n}\Omega. \tag{4.36}$$

Suppose, we want to heat this plasma at power 10 MW. Let us find the needed voltage V to make $P=V^2/R=10\,\text{MW}$. The needed voltage is $V=(RP)^{1/2}$, which will be $(10^7\times 0.204\times 10^{-9})^{1/2}=0.045\,\text{V}$. This voltage is induced by the changing magnetic field through the hole of torus.

Using the Faraday law, $\text{EMF}=\text{Voltage}=-d\Phi/dt$, where Φ is magnetic flux in Webers, so that we need 0.045 Webers/s. Suppose, we have a superconducting coil with radius 1 m, a bit bigger than used in a typical magnetic resonance imaging apparatus, cutting through the donut hole of the torus. Thus, we have $0.045=\pi dB/dt$, or $dB/dt=14.3\,\text{mT/s}$, to provide 10 MW heating of the plasma. If the coil will support 5 T, then the time of the ramping can extend to 348 s. In this time, the energy

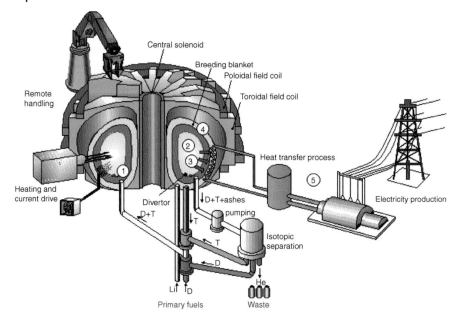

Figure 4.9 Artist's conception (http://www-fusion-magnetique.cea.fr/gb/fusion/principes/principes05.htm) of ITER-based electric power plant. In the figure, (1) marks the injection point of DT mixture into the reaction chamber, where, due to confinement and various heating sources, it goes into the plasma state and burns in fusion reactions (2). At location (3) is marked the release of energy in the form of radiance and fast particles and helium ash. (4) indicates the first wall, of the vacuum chamber, marked "breeding blanket" where fast particles are turned into heat but turn some particles into tritons T that enter the plasma. The first wall, breeding blanket, and vacuum chamber are cooled by a heat extraction system that is used to produce steam and supply a conventional turbine and alternator electricity producing system (5).

$U = 10\ \text{MW}\ 348 = 3.48\ \text{GJ}$ will be available. This is well beyond the 0.207 GJ needed to heat the plasma to 1.5×10^8 K.

One of the difficulties is that the induction heating needs a conductive plasma to start with. So there are other methods used to start the plasma before the inductive heating. These are indicated in Figure 4.9.

Looking at Figure 4.10, note at the bottom that the "Primary fuels" are Li and D. The flow diagram indicates that Li is turned into T, some of which, along with D and ^{4}He, is pumped out at the end of each heating cycle. These residues undergo isotope separation, after which the ^{4}He is emitted as waste and the T and D are reinjected as fuel.

The first role of the breeding blanket is to protect the vacuum vessel and magnets from neutron and gamma radiations. In its breeding function, it produces from lithium the tritium T needed for continued fusion reactions. Finally, it converts neutron energy into heat and transfers that heat to a heat exchanger system. The breeding reaction is based on neutrons from the D–T reaction

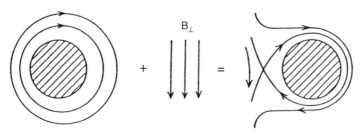

Figure 4.10 Poloidal magnetic field [52] due to the combined toroidal plasma current and vertical field.

$$^6\text{Li} + \text{n} \rightarrow {}^4\text{He} + \text{T} \tag{4.37}$$

$$^7\text{Li} + \text{n} \rightarrow {}^4\text{He} + \text{T} + \text{n}$$

In practice, the blanket can be maintained as a liquid, for example, as a Pb + Li mixture, or as a solid ceramic, such as Li_4SiO_4, Li_2ZrO_3, or Li_2TiO_3.

Figures 4.9 and 4.10 indicate that there is a temperature distribution in the plasma and it is necessary that the first walls be maintained safely far below their melting temperatures. For reference, graphite and tungsten are viable up to the vicinity of 3000 K, but the plasma as we have seen is 150 million K. The heat radial transfer by particle diffusion from the plasma toward the walls is inhibited by the magnetic confinement discussed above. The plasma has no neutral particles, and all charged particles are magnetically confined to move principally in the toroidal direction, around the circle, and this greatly diminishes the ordinary heat flow by diffusion, in the radial direction.

Radiation transfer from the plasma to the walls is important, even though we ignored it in our simple estimate of the necessary heating power, Equation 4.26. Since the DT plasma is simple, containing only electrons and singly positive deuteron and triton positive ions, it does not emit black body radiation as we might suppose. The radiation from the plasma is known as *bremsstrahlung* (acceleration radiation) and can be estimated as follows.

The radiated power density P_{Brem} per cubic meter from *bremsstrahlung* is summarized [51] for a plasma of equal numbers N_D of electrons and ions of charge Z_i. At thermal energy T (in eV), it is

$$P_{Brem} = Z_i^2 N_i N_e T^{1/2} / (7.69 \times 10^{18})^2 \text{ W/m}^3 \tag{4.38}$$

for N_i ions of charge Z_i and N_e electrons. If we evaluate this for the Tokamak plasma, where $Z_i = 1$, with equal numbers of deuterons and electrons, we get

$$P_{Brem} = (10^{20})^2 (12930)^{1/2} / (7.69 \times 10^{18})^2 \text{ W/m}^3 = 19.23 \text{ kW/m}^3. \tag{4.38a}$$

This radiation density P_{Brem} is fortunately quite small compared to the fusion power density for the case we are considering (for the DT case) that is 634 MW/500 m³ = 1268 kW/m³.

This formula makes clear the danger posed by possible W contamination by tungsten walls. Since W has $Z = 74$, with $Z_i^2 = 5476$), entering Equation 4.38 if W ions should be eroded from the wall and end up in the plasma, these ions, even in small amounts, would cool the plasma. This makes graphite an attractive wall material, with Z only 6.

It is not clear what fraction of this radiation might be reflected back by the walls into the plasma, because of its very short wavelength. This energy is probably more likely absorbed by the walls, making it part of the useful power output, but its nature makes it more likely to reduce the temperature of the plasma, than do the fusion events, which actually heat the plasma by thermalization of the trapped energetic charged fusion products.

The distribution of this radiation peaks near $hf = 0.2\,T = 2586$ eV, corresponding to gamma ray photons of wavelength 0.48 nm, where T is the particle energy in eV. The frequency of this emitted acceleration radiation is far above the electron plasma frequency

$$\omega_p = (N_D\, e^2 / m_e \varepsilon_0)^{1/2}, \tag{4.39}$$

which works out to 5.63×10^{11} rad/s for this case, corresponding to 89.7 GHz. The blanket in the Tokamak intercepts this radiation and turns it into heat, protecting the superconducting coils from the damage that the gamma rays might inflict.

For this DT plasma at $10^{20}\,\mathrm{m}^{-3}$, the Debye screening length λ_D, discussed in conjunction with Equation 3.71a, is 84.5 µ. The relation of the Tokamak plasma to the sun and other cases was summarized in Figure 3.18.

It should be pointed out that the power density of the plasma is adjustable from the D,T density. Going from 10^{20} to 2×10^{20} will increase the power by a factor of 4. So, the difference between the full torus volume and the hot plasma volume, perhaps a factor of 2 in effective volume and power output, can easily be recouped by increasing the deuteron density.

4.3.4
Summary, a Correction, and Further Comments

In our desire to make an opaque subject clearer to the nonexpert, we have focused on the essential parts of the nanophysics of fusion. The cross section for electron scattering from the ions, related to the resistivity of the plasma, it turns out, has been treated too simply, and needs a correction factor, which increases the plasma resistance. The correction factor known as "ln(Λ)" multiplies the strong scattering cross section $\pi\, r_2^2$, which when multiplied by T_{Gamow} is the fusion cross section. In the scattering of electrons from ions, the cumulative effect of many small-angle scattering events make this substantial correction. This occurs because the Coulomb force has a long range. The correction does not apply to fusion, but only to the plasma resistance, because in that case the strong scattering is the only way to set up the fusion event, an accumulation of weak scattering events is of no use. The importance of the small scattering events is controlled by the Debye length,

$$\lambda_D = (k_B T \varepsilon_o / n e^2)^{1/2}, \tag{3.71a}$$

which is the maximum distance over which two particles in the plasma experience the Coulomb force. Beyond this distance, the small adjustments of positions of many electrons screen away the direct Coulomb force. In this case, the ratio of the Debye length to the classical turning point spacing r_2, which is called "b" in the plasma literature, is important. Thus,

$$\ln(\Lambda) = \ln(2\lambda_D/b) = \int_b^{2\lambda_D} dr/r. \tag{4.40}$$

Taking $\lambda_D = 84.5\,\mu$ and $r_2 = 111.3\,f$, we find $\ln(\Lambda) = 21.1$, for the DT Tokamak plasma at 150 million K.

This means that the plasma resistivity and resistance values are too small by a factor 21.1, the electron mean free path reduced to 12.18 km, and the plasma resistivity becomes $3.0 \times 10^{-10}\,\Omega\,m$. The voltage to produce the desired 10 MW heating rate has to be increased by a factor $(21.1)^{1/2} = 4.6$. So the needed dB/dt in the example above must be increased from 14.3 mT/s to 65.8 mT/s.

The design for ITER actually employs larger magnetic field values (http://en.wikipedia.org/wiki/ITER), 11.5 T for the toroidal field, leading to a maximum stored magnetic energy 41 GJ, a volume of 840 m³ and a central solenoid field up to 13.5 T (see Figure 4.10), which would limit the time of inductive heating at 10 MW to 13.5 T/65.8 mT/s = 3.4 min, assuming the central solenoid radius of 1 m.

Beyond this, there are several technical issues that in practice are important. The stability of the plasma has been assumed, which is a simplification. The toroidal magnetic field is approximately constant across the cross section, but more accurately decreases from the inner to outer walls of the torus. The induced current in the torus produces the usual circling magnetic field around the current, which is termed a "poloidal" field. In fact, an additional magnet, the "poloidal magnet," produces a vertical magnetic field needed to keep the plasma from expanding outward in radius.

In Figure 4.9, we see that the high-temperature part of the plasma occupies a cross section less than the full vacuum vessel cross section, πb^2, but let us assume $b' \sim \leq b$ is the radius of the current density when the ohmic heating is being applied. (The plasma is actually diffuse and somewhat free to move in the mechanical cross section, and could only roughly be described as a torus of major radius a' and minor radius b'.) The resistance of the plasma R_p (4.36) when corrected for the small-angle scattering factor $\ln(\Lambda) = 21.1$ is raised from 0.204 nΩ to 4.30 nΩ. At the chosen heating power 10 MW, the plasma current I_p must satisfy

$$I_p^2 R_p = 10\,MW, \tag{4.41}$$

which gives $I_p = 48.2$ MA.

One aspect of the plasma equilibrium may be suggested by estimating the magnetic field produced by the plasma current I_p itself. If the current were flowing in a straight solenoid of radius b, the magnetic field $B_p(b)$ at its radius can be gotten from Amperes law,

$$\int \mathbf{B} \cdot d\mathbf{l} = \mu_0 I_p. \tag{4.42}$$

In this situation then, we estimate $B_p(b) = \mu_0 I_p/2\pi b = 9.64$ T, choosing $b = 1$ m. If we imagine bending the solenoid to fit the major radius a, then the field B will become smaller on the outside and bigger on the inside and the relevant factor will be approximately $(a \pm b)/a$. The difference in the plasma-current-induced magnetic fields B_p on the inside versus the outside is then $\Delta B \approx (2b/a) B_p(b)$. If a vertical field $B_z = (b/a) B_p(b)$ is added (to increase the field on the outside), the current I_p will tend to be stabilized, with sum vertical fields equal on the inside and the outside. The required vertical field, in our rough analysis that neglects an outward force from the plasma pressure, is then

$$B_z = \mu_0 I_p/2\pi a = 1.56 \ T \text{ (for } a = 6.2 \ m, \ I_p = 48.2 \ \text{MA)}. \tag{4.43}$$

The vector sum of this circling poloidal field and the toroidal field is then a helical field, which is the field that the ions and electrons in the plasma will follow, depending on I_p.

This situation is sketched in Figure 4.10. On the left, the magnetic field lines shown on a cut through the torus are distorted, stronger on the inside and weaker on the outside, estimated as ΔB above. The uniform vertical field B_z is shown added in the sense to strengthen the field on the outside of the torus. On the right, the resultant field is nearly symmetric as would occur for a straight solenoid. The detailed force balance has to include a component of the plasma pressure to the outside, which arises due to the curvature.

The difference ΔB is the origin of the tendency of the plasma to expand outward, which is corrected by the poloidal coil system. The vertical field B_z, estimated in Equation 4.43, should be applied to maintain the plasma centered in the physical cross section of the torus. A more accurate analysis, which includes the plasma pressure, gives

$$B_z = (\mu_0 \ I_p/4\pi a)[\ln(8a/b) + p_{\text{plasma}}/p_{\text{magnetic}} + \delta], \tag{4.43a}$$

where the correction $\delta < 0$ is given by Miyamoto [52]. Here pressure

$$p_{\text{plasma}} = 2N k_B T = 2 \times 10^{20} \times 12.93 \times 10^3 \times 1.6 \times 10^{-19} = 0.414 \ \text{MPa} = 4.1 \ \text{bar}, \tag{4.44}$$

and p_{magnetic} is of the form $(B')^2/2\mu_0$. If we interpret $B' = \Delta B$ as defined above, then $p_{\text{magnetic}} = (\Delta B)^2/2\mu_0 = 3.84$ MPa. It appears that our rough estimate is within a factor of two of the correct form (4.43a). A full analysis of the equilibrium is given in Section 6.3 of Miyamoto [52].

A variety of undesirable modes have been found to occur and to reduce power. The underlying nanophysics of the fusion power generation is quite clear, but the engineering design of a practical reactor to avoid the plasma instabilities is difficult. Furthermore, there are material degradation problems. We mentioned above the power loss from the plasma if highly charged nuclei such as tungsten are eroded into the plasma. Wall erosion is also a maintenance problem, complicated by the fact that the first wall materials will become radioactive and require special handling.

An analysis of the availability in the sea of deuterium and lithium for a possible but probably unlikely age of fusion-produced electric power has been given by MacKay ([44], p. 172). Seawater contains 33 g of deuterium per ton, and this is a huge supply. Each gram of deuterium represents 100,000 kWh, and the sea contains 197 million tons of seawater per person (at 7.0 billion humans). Lithium is also available in seawater at 0.17 ppm, which translates [44] to 3910 tons of lithium per person. If the lithium produces 2300 kWh per gram in the fusion reactors, and the energy usage per person is 105 kWh/day, then this energy source would last for over a million years. The success of the ITER reactor is not guaranteed, and even so it is not a power reactor but a research reactor. In the words of David MacKay [44], "I think it is reckless to assume that the fusion problem will be cracked."

Even if the problem is cracked, it is clear that a fusion reactor has to be large, a contributor to a power grid. While the fusion reactor seems quite safe, has no meltdown possibilities, and leaves few radioactive waste products, if indeed the engineering of the well-known science into a useful power reactor design succeeds, the end product will be very high technology only possible on a large scale, with high demands on an operator to make it work smoothly.

5
Introduction to Solar Energy Conversion

After learning quite a bit about how the sun's energy is created, and how that process might be reproduced on earth, we turn now to methods of harvesting the energy from the sun. Heat energy is the easiest form to store and distribute and is useful for heating water and home. (The advantage of solar water heating is that the vexing storage and distribution issues are absent in this use of the sun. The saving in electricity from the grid is equally as valid as adding power from solar farms.) The more technical challenge is to turn the sun's energy into electricity, for consumption or perhaps conversion into hydrogen as a portable fuel. Two main tracks of electricity generation are through a heat engine running a generator and direct photovoltaic conversion. An additional track is to use the sun's energy to directly create a fuel, hydrogen, through a photocatalytic water-splitting process that bears some similarity to photosynthesis.

5.1
Sun as an Energy Source, Spectrum on Earth

The main properties of the sun as an energy source are its energy density, 1366 Wm^{-2} at the top of the earth's atmosphere; the highly directional nature of its radiation, and its spectrum. For simple estimates, we will take the sun as a black body at 6000 K.

Starting from the energy density of the radiation given in Chapter 1,

$$u(\nu)\mathrm{d}\nu = [8\pi h\nu^3/c^3][\exp(h\nu/k_\mathrm{B}T)-1]^{-1}\mathrm{d}\nu, \tag{1.1}$$

we multiply by $c/4$ to get a power density, Wm^{-2}, and then convert to wavelength using $c = \lambda\nu$, to get the power density per unit wavelength (as plotted in Figure 1.3):

$$P(\lambda)\mathrm{d}\lambda = [2\pi hc^2/\lambda^5][\exp(hc/\lambda k_\mathrm{B}T)-1]^{-1}\mathrm{d}\lambda. \tag{5.1}$$

As mentioned earlier, the peak in this function is at λ_m, such that $\lambda_\mathrm{m}T = $ constant $= $ 2.9 mm K. The value of $\lambda_\mathrm{m} = 486$ nm and photon energy is 2.55 eV for the solar spectrum in the visible corresponding to $T \approx 5973$ K.

Nanophysics of Solar and Renewable Energy, First Edition. Edward L. Wolf.
© 2012 Wiley-VCH Verlag GmbH & Co. KGaA. Published 2012 by Wiley-VCH Verlag GmbH & Co. KGaA.

The sun's radius $R_s = 0.696 \times 10^6$ km, and its distance D_{es} about 93 million miles (1.496×10^8 km) from earth make its angular radius seen from the earth very small, about $0.266°$. We can infer the total power density at the sun's surface as

$$P = 1366 \text{ W/m}^2 \times (D_{es}/R_s)^2 = 6.312 \times 10^7 \text{ W/m}^2, \tag{5.2}$$

where the geometric ratio of concentration is 46 200. It is this power density that corresponds to $\sigma_{SB}T^4$, with $T = 5973$, where $\sigma_{SB} = 2\pi^5 \, k^4/(15h^3c^2) = 5.67 \times 10^{-8}$ W/m²K⁴.

The concept of the black body radiator is a surface that emits the power density equally into all directions. A hemisphere centered above a black body emitter will receive equal energy on each point of its surface. This can be called a solid angle 2π. (In reverse, a black surface will absorb equal energy from directed beams originating at each point on the hemispherical surface.)

The model for a black body radiator is a small opening into an enclosed cavity at temperature T, such as a pizza oven glowing red on the inside. The opening into the cavity acts as a perfectly black absorber because a light ray entering the small opening has no chance of coming out, independent of the direction it took while entering the cavity. Inside the equilibrium cavity photons propagate in all directions, and any one of these can come out through the hole. The light emitted from the hole is directed equally into each angular range defined by polar and azimuthal angles $d\varphi$ at φ and $d\theta$ at θ. The sun acts as a black body radiator, and many types of surfaces act to a good approximation as black body absorbers and radiators.

Let us imagine a small area of black body surface, with angle from the surface normal taken as θ and azimuthal angle φ. The sun is at a particular angular location and represents an angular diameter of $0.53°$. How can we design an optical system that will maximally concentrate the sun's rays onto this small area of black body surface? The problem is equivalent to designing an optical system that will take all of the light emitted by the black body surface (into all 2π radians) and focus it into the angular range defined by the sun. An example of an optical system that can approach this is illustrated in Figure 5.1.

A concentrator system is perhaps more commonly based on a parabolic mirror as the primary element, and a system of this type has demonstrated a concentration ratio 84 000. Literally, this focused light is more intense than the surface of the sun, and would have intensity $1366 \times 84\,000 = 114.7$ MW/m² if operated above the atmosphere.

The sunlight reaching earth is diminished by specific absorption from molecules in the atmosphere and by Rayleigh scattering of the sort that make the sky blue. Peak illumination at the earth surface is 1000 W/m², in round numbers. The details of the spectrum are more important for semiconductor devices, and for this purpose the spectral power expressed per unit energy is preferable. Figure 5.2 shows the clear day noontime spectrum for the northern hemisphere, a standard spectrum called "AM 1.5," meaning 1.5 air mass traversal, at 48° from the vertical. Most of this intensity is direct light, with a small diffuse contribution from the Rayleigh scattering. Most of this intensity can be concentrated as described above. On a cloudy day, the light intensity may not be too different, but the light certainly is diffuse and cannot be

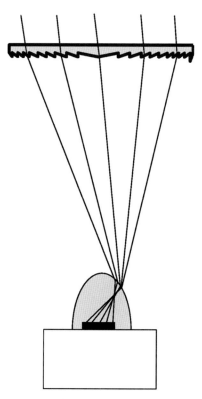

Figure 5.1 Two-stage optical system taking wide angular range emission from hot black surface and focusing it into a small angular range. In reverse, this will focus sun's rays on black surface making use of a large angular range. This system has a primary Fresnel lens and a secondary imaging element made of glass of high index of refraction n. The concentration ratio in such a system can, in principle, reach n^2 $(D_{es}/R_s)^2$ with 84 000 experimentally attained [53].

focused. Focusing systems are almost useless on a cloudy day, while flat solar panels may still operate at 30–40% of their normal capacity.

The direct heat absorption to heat water and to heat a home is actually important in economic terms, perhaps larger than the electric conversion systems that are our primary interest. These systems reduce the load on the electric power grid and should be classified as renewable energy. Building of homes in several countries, notably Israel, requires solar heating used for domestic hot water and to partially heat the home.

5.2
Heat Engines and Thermodynamics, Carnot Efficiency

A heat engine works between a hot reservoir, T_s and an ambient reservoir T_a, and has a limiting efficiency

$$\eta_{\text{Carnot}} = 1 - T_a/T_s. \tag{5.3}$$

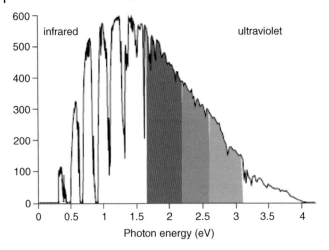

Figure 5.2 Earth-level peak sun's power spectrum, units W/(m² eV). This is the "AM 1.5" spectrum characteristic, appears at noon on a clear day for moderate climates, and 48° from normal incidence. The area under this curve represents 1 kW/m². The shaded areas in range 1.7–3.1 eV represent red, green, and blue parts of the visible spectrum. About half the energy is in wavelengths longer than visible, energies less than 1.7 eV. A small portion of this light intensity comes indirectly by Rayleigh scattering [54].

If we evaluate this at $T_s = 6000$ K for the sun and $T_a = 300$ K for ambient, the value of is 0.95. This analysis can be extended to better simulate a solar thermal power plant. In such a case, a focused light system puts sunlight on an intermediate collector heat reservoir, which rises to an intermediate temperature T_c. The collector reservoir then is the hot reservoir for the heat engine. In this case, radiation back from the intermediate collector to the sun is an important part of the analysis. In Figure 5.1, one can imagine the intermediate reservoir being the black surface at the bottom of the field, which we now say will come to temperature T_c and radiate a power density $\sigma_{SB} T_c^4$ back to the sun through the optical system, which is a Fresnel lens in Figure 5.1. In this circumstance, which was investigated by Landsberg, the efficiency is expressed as

$$\eta_L = (1 - T_c^4/T_s^4)(1 - T_a/T_c) - \delta. \tag{5.3a}$$

The maximum value of this efficiency, $\eta_L = 0.854$, setting the correction δ to its minimum, zero, with $T_s = 6000$ K and $T_a = 300$ K, occurs at collector temperature $T_c = 2544$ K. The terms in this expression clearly relate to radiation equilibrium between the sun and the collector, and to Carnot efficiency of the heat engine working between the collector at T_c and the ambient exhaust temperature T_a. The correction δ is related to entropy generation, $\delta = T_a \, (dS/dt)/P_s$, where P is the power from the sunlight. Entropy $dS = dQ/T$ is accepted from the sun by black body absorption, some is returned to the sun by the black body emission from the collector, and a contribution $dS_c/dt = P_c/T_c$ is created by black body radiation from the collector to the Carnot engine.

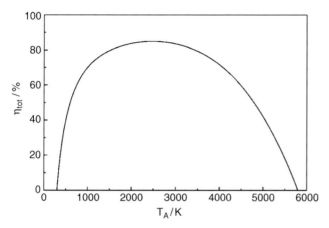

Figure 5.3 Plot of system efficiency η_L (mechanical power output divided by sun power input) versus collector temperature (referred to as T_c in text), assuming sun at 6000 K and ambient exhaust at 300 K. The peak efficiency is 0.854 at 2544 K ([53], p. 52, Figure 3.1.

A plot of the efficiency as a function of the collector temperature T_c is shown in Figure 5.3.

5.3
Solar Thermal Electric Power

This category is also known as concentrated solar power, CSP, with two main forms, the large central receiver and the parabolic trough system. Large central receiver solar thermal electric power plants are shown in Figure 5.4. Design of the mirrors and their tracking is a feat of engineering. Only in a large system does it seem likely that the theoretical efficiency 84.5% could be approached because of the difficulty in maintaining components and heat-carrying fluids at temperatures approaching 2544 K. In such high temperatures, it seems that molten salts, or a molten metal such as lithium, with boiling point 1615 K, are good candidates, because oils will decompose and steam would require an unattainably large pressure. From Figure 5.3, Equation 5.3a, the efficiency at 1615 K is 0.81, much higher than found in photovoltaic systems, as we will see. In fact, the PS10 system, left in Figure 5.4, has 11 MW capacity, with 624 metal glass mirrors of 120 m^2 area, and uses pressurized steam at 250 °C and 40 bar (http://www.solarpaces.org/Tasks/Task1/ps10.htm.).

So, this system, working at 250 C, must be far short of the available efficiency. As a benchmark, tungsten has a melting point 3683 K, has been used in light bulbs for decades, and would allow a much higher temperature receiver assembly. Other high-temperature metals are molybdenum, tantalum, and titanium. (Aircraft manufacturers and architects have learned to use titanium to considerable advantage.)

At 250 °C, the theoretical system efficiency is in the vicinity of 43%, by looking at Figure 5.3 and using Equation 5.3a. An empirical efficiency estimate for PS10 can be made easily: the receiving area is 624 × 120 m^2, with input 1000 W/m^2,

Figure 5.4 Advanced central receiver thermal solar power installations PS10 (*left*) and PS20, near Seville, Spain. These systems of accurately tracking mirrors (functionally similar to Figure 5.1) image the sun onto the central receivers, which are described as the collector at temperature T_c in Equation 5.3. Following the analysis above, maximum system efficiency η_L (mechanical power output divided by sun power input) will occur at $T_c = 2544$ K. Achieving such a high receiver temperature is an as yet unmet challenge in materials science and engineering (http://www.solarpaces.org/Tasks/Task1/ps10.htm.).

so $P_s = 74.9$ MW. At 11 MW rated capacity, the efficiency is 14.7%. This is a situation, unlike that in silicon solar cells, where a large increase in efficiency is potentially available. The operating efficiency of this plant has also been given as 18%.

In practical terms, since the sun's input is available at no cost, the advantage of a more efficient system is directly in the available power, which could be raised from 11 MW (at 14.7% efficiency) to 60.7 MW at 81% efficiency (molten lithium heat fluid, assuming 300 K ambient).

To estimate the value that would come from the added 49.7 MW potential capacity, let us assume all the power is put into the electric grid at 0.1 \$/kWh. On a yearly basis, the added revenue from the upgrade, assuming the sun power is available 0.33 of the time, would be $0.33 \times 49.7 \text{ MW} \times 3.15 \times 10^7 \times 0.1\$/3.6 \times 10^6 = \$14.4$ million/year. This seems a large payoff for upgrading a portion of the system. (The land area and the mirrors would be unchanged in this upgrade.) The reason that this was not done may be that the technology for the (tungsten receiver/molten lithium heat fluid) system is not available, but one can see potential for such an advance. The design of a turbine that will convert the high temperature into mechanical energy is also an aspect for possible improvement.

In European Community report dated 2006 (http://ec.europa.eu/energy/res/sectors/doc/csp/ps10_final_report.pdf.), it is stated that the total investment in PS10 is 35 million Euros, about \$52.5 million. In a crude assessment, the cost per watt is

$52.5 M/11 MW = \$4.77/W$. This report (http://ec.europa.eu/energy/res/sectors/doc/csp/ps10_final_report.pdf.) makes it clear that the location was optimum (see also Figure 1.7a) both from the point of view of available sunlight and from the point of view of the cost of land.

This type of installation can be improved to reach its potential high efficiency with high technology in the heat collector and heat fluid and turbine design. An example of a smaller but higher performance system is shown in Figure 5.5. A higher operating temperature is available in a closed cycle Stirling engine using helium or hydrogen gas, with no need for a heat fluid transport. In a Stirling engine, the same limited volume of gas cycles back and forth in a repetitive fashion.

A report of $\eta_L = 0.31 = P_{electric}/P_s$ efficiency in a separate Stirling engine system (https://share.sandia.gov/news/resources/releases/2008/solargrid.html.) indicates that a substantially higher collector temperature is available in Stirling engines. The reported dish had 82 mirrors focused on a 7 in. aperture of the hydrogen-based Stirling engine. The 31% efficiency was demonstrated as electric power output divided by solar energy input, and the record was set on a cold day such that the presumably air-cooled cold port of the Stirling engine could be maintained at $23\,°C = 296\,K$. Over a period of several hours during the day, the dish system maintained an electrical output 26.5 kW or about 36 hp. A substantial exhaust heat

Figure 5.5 EuroDISH parabolic mirror Stirling engine on tracking platform developed by Plataforma Solar de Almeria (PSA) in Spain. A closed cycle of hydrogen or helium gas in the compact Stirling engine unit allows a higher collector temperature, increasing the efficiency to ~30%. This is a self-sufficient power source that might be used in a rural location, augmented by electrical storage in batteries or by electrolytically produced hydrogen from water, that could be used on demand to run fuel cells (http://en.wikipedia.org/wiki/File:EuroDishSBP_front.jpg.).

flow is also generated at $23\,°C = 73\,°F$, evidently $P_{exhaust} = 0.69 \times 26.5\,kW/0.31 = 59\,kW$. If water-cooling was employed, a substantial supply of water at $73\,°F$ on a winter day might be useful for home heating, but typically such exhaust energy is wasted.

All thermal solar power systems, unlike nonconcentrating photovoltaics, require cooling. In practice for large plants, this means that water is needed. It is noted that in California the use of drinking water for this purpose is illegal, suggesting that common practice is to dump the cooling water.

Smaller installations based on Stirling engines heated by plastic Fresnel lenses are possible, for example, for a remote household power supply, typically to charge batteries. An acrylic Fresnel lens of 0.5 m diameter and 2 mm thickness is inexpensive, and it is possible to set up a Stirling engine so that fossil fuel may alternatively heat the Stirling intake when the sun is not shining. Based on the performance noted above, peak sun power would, at 30% efficiency, generate at best 60 W. While this might be more expensive than a larger area of photovoltaic cells, the latter would not allow fossil fuel backup.

5.4
Generations of Photovoltaic Solar Cells

The single junction solar cell was described in Chapter 3, before Figure 3.17. The limiting efficiency of a single junction cell is 31% at one sun illumination, but as high as 40.8% with concentrated sunlight. The Si single-crystal solar cell shown in Figure 3.17 is the most common first-generation solar cell. Elaborations on this cell reaching 24% efficiency will be described in Chapter 6, but the bulk of the Si cells are characterized, as in Figure 5.5, in a range 15–20%. The crystalline Si solar cell is built on a wafer of Si, which is typically 200–300 μ thick, cut from a large crystal using a wire saw that wastes also one wire-width of the Si single-crystal boule (http://www.omron-semi-pv.eu/en/wafer-based-pv/wafer-preparation/slicing-the-ingot.html.).

The cost of chip-based cells is correspondingly high, here estimated in the range of \$3.50/peak watt.

An influential summary and prediction of the costs in solar cell types, dating from 2003, is shown in Figure 5.6. This figure also shows the fundamental limits on the efficiency of cells. It is clear that a thin-film solar cell structure noted as type II in the figure will cost less, because the thickness of the active portion of the cell is reduced to a few microns, limited by the absorption of light in the material in question. Thin films of silicon can be deposited by several methods, representing the most common form of thin-film type II cell, shown as type II in Figure 5.6. This class of single junction cell is here characterized at \$1/W with efficiency in the vicinity of 10%. These devices are inherently less expensive than those based on single crystals of Si because they can be deposited in depths of only a few μm on wide area substrates of glass or metal foil. There are large continuing markets for smaller scale cells or panels where the efficiency is less important than price, such as watches, calculators, roofing tiles, power for roadside signs and telephones, "see-through" amorphous panels for windows, or auto sun-roofs, which have been served by inexpensive amorphous

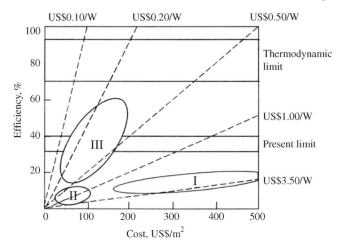

Figure 5.6 Efficiency–cost trade-offs for three generations of photovoltaic solar cells, cost in 2003 US dollars. The categories are I wafers, II thin films, and III "advanced thin films" [55].

silicon solar cells. A related form of inexpensive silicon thin-film cell is actually a tandem cell with three different bandgaps based on amorphous Si, semicrystalline Si, and Si–Ge alloys. However, the efficiency of these devices has not exceeded 10% and they are best characterized as type II rather than type III. By far the most common solar cell is silicon in type I or type II form, followed by CdTe thin-film cells.

The goal of "grid parity" describes the effort to find solar cell systems cost competitive with coal- or gas-fired turbine electric power. A consensus view is that silicon-based products will not reach this goal: crystalline on the basis of cost and thin films on the basis of efficiency. (There may be an opening for crystalline Si cells in concentrating systems, however, as we mention later in this chapter.) The "grid parity" goal assumes the power can be fed into an existing power grid, which avoids the basic problems of intermittency of sunlight, and variability of power under weather conditions. A stand-alone solar system would have to provide storage in massive amounts or connection to a grid so large that the sun is always shining on some part of it. Storage in a totally solar power grid might be provided by diversion of some of the daylight power to hydrogen electrolysis and fuel cell capacity to meet power demand overnight. A stand-alone renewable power system might alternatively be a combination of solar and wind farms, with the lower power usage at night provided by wind, using the solar part to bring the total power up to peak demand in the day.

The leading contenders for "grid parity" systems appear to be type II thin-film cells of CdTe, followed by CIGS cells (copper indium gallium selenium). The First Solar firm, leader in production of CdTe thin-film cells, has a signed agreement with China to build a 2 GW solar cell farm in Ordos City in the Mongolian desert [56].

The second leading type II thin-film cell is CIGS, copper indium gallium selenide. CdTe and CIGS thin-film cells are both more efficient and cheaper to produce than earlier thin-film Si cells. To what extent they are limited by the relatively small supply of ingredients such as Te, In, and Ga, and the toxicity of Cd, is not entirely clear.

The type III cell, at least in thin-film form, is purely conceptual at the present date. It appears that the only cells to operate in or above the limiting range between 30 and 40% efficiency are, in fact, multibandgap epitaxial structures, containing crystalline layers of GaAs, AlAs, InAs, and germanium. These cells are so expensive that they are credible for large-scale application only in conjunction with concentration factors of several hundred. A mosaic of such cells would be placed, for example, at the focus of a parabolic reflector as shown in Figure 5.5, and would definitely need water-cooling to keep the expensive cells in a safe operating temperature range. The multi junction concentrating cell systems, to be described later, appear to be more expensive than the 0.2–0.4 \$/W suggested in Figure 5.6 as typical of type III cells. Yet, large-scale concentrating multijunction systems have been presented as capable of "grid parity" in favorable locations such as the American southwest or southern Spain. Although such type III concentrating systems are proposed as competitive with large arrays of simpler type II cells such as CdTe or CIGS, the fact is that no large facility to date has been based on the light-concentrating multijunction cell systems. The marketplace decision to date may be affected by the higher complexity of the concentrating system compared to a large array of thin-film modules, which do not require tracking and water-cooling.

The balance on this choice may yet change. The concentrating systems clearly minimize the volume of semiconductor materials, notably do not require large amounts of Cd, Te, In, or Se. While performance has been lower, large-scale concentrating systems based on crystalline silicon (of wide abundance) may yet win over the multijunction systems, which are based on GaAs, AlAs, InAs, and germanium and have reached 41% efficiency. In principle, a single-bandgap cell, for example, crystalline silicon, with concentration by mirrors or lenses, can reach 40% efficiency. The multijunction cells are a proven technology, widely deployed in space programs going back to the Sputnik Russian space program era. The technology is available, but its cost and complexity are primary issues, ahead of possible material supply issues, in adapting to solar farms on earth.

Figure 5.6 suggests that type III cell cost might, in principle, be as low as 0.2 \$/W, although there seem in 2011 to be no examples of thin-film cells above 20% efficiency. Martin Green, leading researcher and author [55], stated, "There would be an enormous impact on the economics if these new (Type III) concepts could be implemented in thin film form, making photovoltaics one of the cheapest known options for future energy production." "Third generation" is an appealing description for any manufacturer who wants to sell his latest product. Following Martin Green, Figure 5.6, we will reserve the terms "Type III" or "Third Generation," to devices in the 20–60% efficiency range. The only existing type III devices, according to this definition, are not thin films but concentrating epitaxial crystalline multijunction devices based on GaAs molecular beam epitaxy, liquid phase epitaxy, or organometallic epitaxy.

We will return to the physics of the efficiency limits in the range 30–40% for single junction cells in Chapter 6.

5.5
Utilizing Solar Power with Photovoltaics: the Rooftops of New York versus Space Satellites

One of the questions on many aspects of solar and renewable energy is the optimum scale of the device or system. We have seen in Figures 5.4 and 5.5 concentrating solar thermal installations on the scale of 11 MW and 25 kW, respectively, with a good indication that the present efficiency of the smaller installation is nearly twice that of the large installation. (It is not clear what the cost in $/W comparison is, but it probably favors the large system.) It has been argued above that much higher efficiency than the present 15% should be available in the solar tower systems with advances in engineering and in materials. We will see later that arrays of small concentrating solar cells built into panels are commercially available using small plastic Fresnel lenses (as suggested in Figure 5.1) and in a miniature form of the parabolic reflector as shown in Figure 5.5. A long-standing suggestion has been for concentrating arrays above the earth in space, with transmission of power to earth by microwaves. The cost has so far been prohibitive. A large but small mass parabolic mirror in space might be constructed cheaply from aluminum foil, or aluminized mylar, to be unfurled after reaching orbit. The positioning of such power satellites above large cities would avoid the difficulty of building new power lines on earth. In principle, an array of power satellites, following the lead of the GPS satellites, might provide power 24 h. Many large cities are coastal or on large lakes, where the receiving units could be located on barge arrays with underwater DC power lines to land.

The City of New York has recently carried out a careful mapping assessment of the solar power capacity of its one-million-plus rooftops [57], concluding that 66.4% of the buildings of the city had roof space suitable for solar panels. It was concluded that 5.847 GW could be generated putting solar panels on hundreds of thousands of buildings. It was concluded that 49.7% of the city's peak power usage could be generated from solar power and 14.7% of the city's annual electricity use, taking into account typical weather conditions. David Bragdon, director of the Mayor's Office of Long-Term Planning and Sustainability said the city could realistically add "thousands of megawatts" in solar power. The city would likely establish a uniform approach and presumably would negotiate with a company like First Solar to do the installations. The budget of the city is $65.7 billion for 2012. Nominally, the installation would cost around $5.8 billion that over 10 years would be 0.9% per year of the city budget.

Such a project would be a large-scale deployment of small installations, and would be in the power of the city to mandate. The article states that U.S. nationwide installed solar capacity is 2.3 GW, less than half the rooftop potential of New York City. We will return in Chapter 11 to some of the issues regarding the deployment of renewable energy resources, which on the national level is made difficult by the sway of vested fossil fuel interests over the Congress. The existence of the Mayor's Office of Long-Term Planning and Sustainability, able to obtain city funding of $450 000 for the aerial survey, is in itself encouraging, and suggests that large cities may be an avenue toward rational implementation of renewable energy systems.

5.6
The Possibility of Space-Based Solar Power

The idea of putting solar converters in space, and beaming the power to reception areas on earth, by microwaves or by lasers has been discussed [58] for decades. One possible scheme, based on concentrating mirrors, a "heliostat" in space, is sketched in Figure 5.7 [59].

The conceptual heliostat design of Figure 5.7 (one of the five designs proposed in the cited report) is envisioned to be stationary at 22 700 miles above a location on earth. Tilting of the primary and secondary mirrors optimizes power collected by the energy converter during the 24 h cycle. The size of this system is large: to make a suitably collimated beam at 5.8 GHz, the satellite transmitting antenna would have to be about 500 m in diameter, powered by a phased array transmitter, and would require a 7.5 km diameter rectifying antenna "rectenna" on earth. The power density at this receiving antenna would not be dangerous, but this expensive system would provide only 1.2 GW on earth. The system would have to be assembled in space, as was the International Space Station (ISS), which, in a Low Earth Orbit, LEO, cost $130 billion. This would suggest more than $100/W cost, 100 times larger than on-earth solar or wind installations. Electric power usage in America is on the order of 500 GW, so the Heliostat Power Satellite device indicated in Figure 5.7 would be a relatively

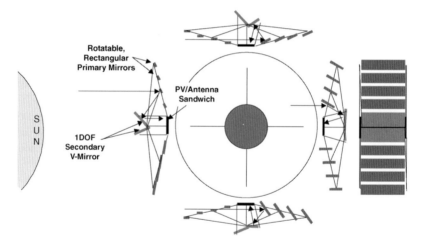

Figure 5.7 Heliostat conceived for geosynchronous earth orbit [59]. A sketched system of focusing mirrors, photovoltaic or thermal energy converters, and transmitter(s) to earth, placed in geosynchronous earth orbit at altitude 36 000 km (22 370 mi). In Ref. [59], this figure "1DOF" means "one degree of freedom." One such power satellite could equally serve any North American city, providing power almost continuously, but requiring a dedicated rectifying antenna of 7.5 km diameter on earth, assuming a 500 m phased array transmitting antenna on the satellite, and delivering 1.2 GW on earth. The cost of this system is likely to be more than $100 billion, corresponding to $100/W. The transmission frequency is 5.8 GHz, corresponding to wavelength 5.2 cm. This band transmits well through clouds and even light rain.

small contribution. A prior power satellite design of a generally similar nature [60] called for a 15 km linear array of 344 combined solar collectors/converters, including power cables up to 15 km in length. The power satellite was to be launched in 340 segments, each containing rotatable solar arrays and part of the transmitter, which would be assembled in orbit using on-segment electrical propulsion. Economical earth to orbit transportation capable of 30 ton payloads would be needed to assemble one such satellite. A set of smaller geosynchronous satellites, adding up to 1.2 GW would be possible with laser transmission of energy to earth, avoiding the large transmitting antenna (http://science.nasa.gov/science-news/science-at-nasa/2001/ast13nov_1/.). It appears that the reception area for each laser-based satellite, a field of solar cells, would require the same area, 7.5 km in diameter. This hypothetical laser version has the further drawback that clouds would block the power, unlike the microwave version that will operate in cloudy weather. It is clear that these designs are all extremely expensive, and likely unsuitable for an urban power supply, on the basis of the 7.5 km diameter receiving space requirement.

The reports cited are nearly a decade old and do not reflect experience from the International Space Station (http://en.wikipedia.org/wiki/International_Space_Station.). ISS is at altitude varying from 400 to 335 km, maintained by onboard power, and has been manned since the late 2000s and projected to extend at least to 2020. It has 262 400 solar cells (http://science.nasa.gov/science-news/science-at-nasa/2001/ast13nov_1/.) located on 8 arrays of dimension 34 m × 11 m, cell area about 2500 m^2, and delivers about 110 kW. NiH batteries are charged during illumination to provide power during the 36 min of each 92 min orbit in which the sun is obscured by the earth. The cost of the ISS space station has been estimated as $130 billion over 30 years.

The cited studies generally consider neither thin-film solar cells nor the possibility of unfurling thin aluminized mylar mirror surfaces, such as the 10 m^2 "nanosail" (http://science.nasa.gov/science-news/science-at-nasa/2011/24jan_solarsail/.), which might be the basis for a less massive structure along the lines of Figure 5.7.

Progress has been made in the private space industry over the past decade. A recent event is the award, in June 2010, of a $492 million contract (http://www.freerepublic.com/focus/f-chat/2537487/posts.) to SpaceX, by Iridium Communications, to launch tens of communications satellites at 666 km altitude, as part of Iridium's $2.9 billion plan for its upgraded communication satellite network, called NEXT (http://en.wikipedia.org/wiki/Iridium_satellite_constellation.). This work will in part use the SpaceX Falcon 9 launch vehicle, capable of putting 11 tons in low earth orbit.

The Iridium NEXT network to be launched starting in 2015 will deploy 66 communications satellites at 666 km altitude, plus 6 in-orbit and 6 on-ground spare satellites, approximately 1000 kg in mass. The price $2.9 billion for an in-orbit fleet of 66 low earth orbit satellites is much below the ISS cost estimated as $130 billion. We can think of adapting satellites of this type, perhaps outfitted with parabolic mirrors to be unfurled in space, concentrating solar cells, and phased array microwave transmitters to send power to earth.

For a solar power satellite in 666 km orbit, if we accept the 7.5 km needed rectenna size from 36 000 km, one can estimate the required rectenna size by linear scaling. Using the same 500 m diameter phased array antenna, described for the power satellite of Figure 5.7, one would need a (666 km/36 000 km) × 7.5 km = 139 m diameter dedicated ground "rectenna." This area is equivalent to 1.89 football fields or 3.75 acres, a reasonable size for urban satellite power system, small enough to be cordoned off to allow a higher microwave power density. A rectenna is an array of dipole antennas with diodes mounted to produce direct current. The spacing of the many elements would be on the order of the wavelength, 5.2 cm. This array could be elevated above ground so that the ground beneath would receive sunlight (possibly could have solar cells) with near-zero microwave power density at ground level. Admissible legal occupational 5.8 GHz power density is 50 W/m^2, while the power density to deliver 1 GW to the proposed area is 66 kW/m^2. As we know, full sunlight is about 1 kW/m^2 and full power inside a typical microwave oven is 10 kW/m^2. The envisioned beam would be deadly, 6.6 times more intense than a microwave oven.

On this basis, it seems unlikely that such a satellite power system could be implemented near New York, although the receiver area would be more dangerous but cover less area than tolerated hazards of miles subway tracks, heliports, airport runways, and freeways. A reasonable microwave power density, equivalent to full sunlight, 1 kW/m^2, would deliver only 15.2 MW, which seems not enough to be useful. (The Indian Point Nuclear Reactor 40 miles north of New York has three reactors totaling 1.9 GW.) From the solar power satellite, one GW could be delivered at the 1 kW/m^2 intensity level if the beam were split equally onto 66 areas of the 3.75 acre size. A phased array antenna on the solar power satellite could, in principle, address 66 separate rectennas in rapid sequence. This might offer also a solution to the problem of building more power lines, as part of the distribution can be done directly from the satellite.

We can make a more detailed model for a power satellite "constellation" in sun-synchronous orbit, extrapolating from the Radarsat-1 and -2 satellites developed and launched by the Canadian Space Agency, with 100.7 min orbital period, total cost $1.145 billion. Radarsat 1 (mass 2750 kg) was launched in 1995 and has completed 15 years of data collection in a sun-synchronous orbit near 807 km altitude. As suggested in Figure 5.8 (Radarsat-2, 2200 kg, launched in 2007), these satellites access a 1000 km width of land along their path with a synthetic aperture radar (SAR), which uses a phased array of transmitting elements. The high-resolution mode addresses at minimum a swath 45 km in width, larger than the 7.5 km rectenna envisioned for the geosynchronous earth orbit (GEO) satellite of Figure 5.7. The "SAR" antenna elements are stationary and the directionality is accomplished by timing or phasing of the signals from those elements. The satellite always faces the sun at approximately the same angle, and the advantage, if this were turned into a solar power satellite, is that tracking of the sun to optimize power would be facilitated.

The satellite does 14 7/24 orbits in its fixed plane as the earth rotates once underneath it, in 24 h. A small effect associated with the particular angle 98.594° tilts the orbital plane a tiny amount each day so that during the course of the year the

5.6 The Possibility of Space-Based Solar Power

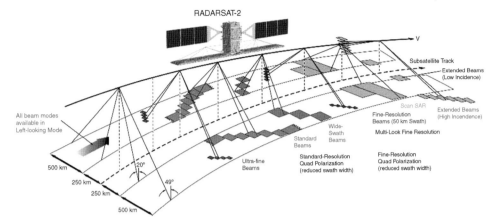

Figure 5.8 (Radarsat-2, 2200 kg, launched in 2007) Imaging capabilities [61] of Radarsat-2. The synthetic aperture radar antenna is the rectangular slab at the bottom (15 m × 1.5 m, mass 750 kg), an array of C-band (5.405 GHz) transmit and receive modules. The solar arrays are each 3.73 m × 1.8 m. The full cycle of orbits repeats after 24 days, with 14 7/24 orbits each 24 h. The orbit is nearly polar (within 8.6°), with 100.7 min period. The inclination of the orbital plane is 98.594° from the equatorial plane. The Radarsat track presumably goes over Ottawa, Canada, say at noon, and this condition repeats at Ottawa every 24 days. A constellation of 24 such satellites could put one satellite above Ottawa at noon each day.

same side of the satellite continues to face the sun. This is a "sun-synchronous" orbit, a special type of polar orbit.

All of the orbits go through points close to the north and south poles, and the orbits are most widely spaced, about 2860 km ~ 1780 mi, at the equator. At 40 °N latitude, this spacing is 1363 mi.

One can imagine a power satellite version of the Radarsat-2 satellite whose orbit passes over New York City. Distinct rectenna locations within a swath of 1000 km = 621.5 mi width could then be provided power, while the satellite is overhead. The antenna would have to be larger (our previous estimate was 500 m diameter) to reduce the minimum swath from 45 km, to focus only on our chosen rectenna area, 139 m in diameter. The maximum Radarsat-2 power is 5 kW, and the antenna mass is 750 kg. The Radarsat antenna is 15 m × 1.447 m = 21.705 m^2, and contains 10 240 radiating elements [61], which are organized in 2 wings, each with 2 antenna panels. If we assume that each radiating element is $\frac{1}{2}$ wavelength, spaced longitudinally by $\lambda/2$, and columns of elements are spaced laterally by 1 wavelength λ, then the area of the antenna would be $10\,240 \times \lambda^2 = 10\,240 \times (0.052)^2 = 27.7\,\text{m}^2$, which is similar to the stated area, 21.705 m^2. The whole satellite is rotated along its direction of motion, in a roll maneuver taking 10 min, to switch from a right view to a left view. So the rapidly accessible addresses on the ground are in a single 500 km swath on one side or the other of the orbital ground track.

But the time above a given location turns out to be insufficient for useful power delivery, on this model. If we take the longitudinal addressable length as the satellite altitude, $A = 807$ km (which would allow angles up to 45 ° from the vertical), then the useful time it can send power to a single location on the ground is

$$\Delta T = TA/2\pi R_e = 100.7 \text{ min} \times (807/2\pi R_e) = 2.034 \text{ min}, \tag{5.4}$$

with T the orbital period, $R_o = R_e + A = 7164$ km, altitude $A = 807$ km [61], taking $R_e = 6357$ km for the polar radius of the earth. $\Delta T = 2.034$ min is clearly too short a time for useful power transfer, so the orbit should be adjusted to a higher altitude. This will require a larger antenna to keep the same resolution on the ground. To get a useful time, let us choose 2.034 h, 60 times longer. We can find the new orbit radius $R'_o = R >_e + A'$ to make the period $\Delta T' = 60 \times 2.034$ min $= 2.034$ h by using Kepler's law

$$(T)^2/(R_o)^3 = \text{constant}. \tag{5.5}$$

To find the new parameters T', A', R'_o such that $\Delta T'/\Delta T = 60$, setting $R'_o/R_o = x$, and $R_e/R_o = a = 0.887$ we take the ratio based on Equation 5.4, and using Equation 5.5:

$$\Delta T'/\Delta T = 60 = (T'/T)(A'/A) = x^{3/2}[(x-a)/(1-a)] = x^{3/2}(x-0.887)/0.113, \text{ or} \tag{5.6}$$

$$6.78 = x^{3/2}(x-0.887). \tag{5.6a}$$

The solution to this numerical equation is $x \approx 2.55$, corresponding to orbit radius $R'_o = 2.875\ R_e$, $A' = 1.875\ R_e = 11\ 918$ km, and period $T' = 6.83$ h. These numbers return $\Delta T' = 2.03$ h using Equation 5.4. (A slight change to make the period 6.0 h will be assumed below, to give exactly four orbits per day.)

We are pursuing a design of a solar power satellite based on scaling the properties of the Radarsat, and have found that to make the time the satellite is in view of a particular location long enough to transfer a significant amount of energy, we have had to make altitude $A = 11\ 918$ km versus the 807 km, an increase by 14.8. (This altitude is about 1/3 of the geosynchronous orbit at 36 300 km mentioned in Figure 5.7.) The new orbit is not assured of being sun-synchronous, but we will assume that it is [62], or can be adjusted to become sun-synchronous, (For example, if the period T' is adjusted to 6 h from 6.83 h, then the satellite will make exactly four earth orbits per day.) The sun will always be approximately at right angles to the track of the satellite, so that the vertical solar cell panels in Figure 5.8 will be reasonably oriented to absorb light from the sun year-round. In the larger orbit we have assumed, there will be no period of eclipse, realizing that the orbit radius is now 2.9 earth radii. The width of the ground that can be accessed is about $A\ [R_e/(A + R_e)] = 0.655\ R_e = 0.655 \times 6357$ km $= 2600$ mi. If the satellite is above New York, its power could be tapped in locations 1300 miles to the west, which would include Chicago and a bit more. Even so, we are insisting on resolution to restrict the incoming energy to a receiving region 1309 m in diameter. This requirement will make the transmitting antenna large in dimension, but still could contain only the 10 240 elements in the Radarsat, the elements would be more widely spaced.

5.6 The Possibility of Space-Based Solar Power

These are generous assumptions and we use them to judge the feasibility of a power satellite to deliver 1 GW to earth. The solar array will have to be enlarged to provide 1.5 GW on the satellite, and if we assume the cells are 40% efficient (at present possible only with advanced cells using mirrors for light concentration), the cells will need to intercept 1.5 GW/0.4 = 3.75 GW. At 1.366 kW/m² the area of the solar collector is 2.745×10^6 m², or a square area whose side L is 1656 m (about 1 mile). This large solar array could be assembled in orbit along the lines of the International Space Station, as suggested by Figure 5.7. If the cells are in thin-film form, they might conceivably be unrolled once in orbit, making use of weightlessness. The weightlessness in orbit means that the supporting structure for the arrays of solar cells and antenna elements need not be strong, and thus they need not add a lot of mass. But the overall mass of the cells, antenna elements, and the transmitter and cabling will still be large and make it expensive to put the system into orbit. Let us make minimal estimates of the masses that would be required.

The working part of a thin-film solar cell needs only to be a few micrometers in thickness. A common commercial thin film is $1/2$ mil mylar, aluminized to form a "space blanket" that reflects heat back to keep the person warm. If our solar cells are 10 μm thick of semiconductor and conducting metal, plus $1/2$ mil of mylar, then the overall thickness is 10^{-5} m + 0.5 (2.54 cm/1000) × 0.01 m = 2.27×10^{-5} m. If we assume this thin-film solar cell to have an average density of aluminum, 2700 kg/m³, then the solar panel as modeled, to deliver 1.5 GW, will have mass

$$M = 2700 \times 2.27 \times 10^{-5} \text{ m} \, 2.745 \times 10^6 \text{ m}^2 = 168\,000 \text{ kg}.$$

At 907 kg per U.S. ton, this is 185 tons of solar collector. The International Space Station mass is 417 289 kg, is only about 2.5 times bigger. This mass is unavoidable and cannot be reduced, to intercept 3.75 GW of sunlight.

The antenna mass of Radarsat is given as 750 kg, and is 1.5 m wide. The power satellite antenna width will be much larger, although it may not need a larger number of elements, for two reasons. The altitude ratio is 14.8 as noted above, and the desired resolution on the ground is to be reduced to 139 m from 45 km, this is a ratio 323, so the total width enlargement factor is 323 × 14.8 = 4791, and the new width of the transmitting antenna is 7186 km. This is extraordinarily large, but could be thought of as an array of widely spaced dipole elements: the overall size, not the density or weight, is the determining factor. As the second estimate, we extrapolate from 500 m antenna for GEO (Figure 5.7) at 36 000 km altitude by 1/3 for altitude 12 000 km, and multiply by 7.5 km/139 m (since we require the rectenna to occupy only two football fields, versus 7.5 km), we get an estimate for antenna dimension = (500/3) (7500/139) = 8992 m, fairly close to the earlier estimate of 7186 km. These two estimates are quite close, and impractically large.

To finish our projection of the Radarsat to make a 1 GW power satellite, let us take the new dimension as 8000 m, compared to the 1.5 m antenna width on Radarsat. If we were to linearly scale the mass of the antenna (750 kg on Radarsat), we get an antenna mass estimate 4×10^6 kg, or 10 times the mass of the International Space Station. It seems likely that the transmitter mass is more likely the right quantity to scale, since the antenna can be obtained at nearly constant mass just by spacing more widely a fixed number of elements. Although the transmitter element mass is not known to us, it may well amount to the 4×10^6 kg mass we just obtained. This rough estimate can be presented as 1 GW/$(4 \times 10^6 + 168\,000)$ kg = 240 W/kg.

For comparison, an estimate of the mass of a 4 GW space station as 80 000 tons = 80×10^6 kg has been presented (http://en.wikipedia.org/wiki/Space-based_solar_power.) corresponding to 50 W/kg. This mass is larger in part because it assumed conventional solar panels, not thin-film solar cells.

Our estimates suggest that the mass of one power satellite to deliver 1 GW to areas on earth, the size of two football fields, is at least 10 times the mass of the International Space Station. If we take the cost as proportional to mass, then the cost of the proposed 1 GW power satellite is at least 10 times the cost of the ISS, $10 \times \$135$ billion, with cost at least \$1350/W. This is a thousand times more expensive than solar power on earth, wind power on earth, or a conventional coal- or gas-powered electrical plant. In fact, a constellation of at least four such satellites would be needed to extend the 2 h above a given location to the peak power of the working day. If four were in orbit, then any part of the country could obtain power but the total power available would still only be 4 GW.

From an entrepreneurial viewpoint, there may be organizations willing to pay much higher prices for reliable electrical power literally at the ends of the earth. Power could be in this way reliably be provided at the poles of the earth, anywhere at sea, or in remote mountainous locations to support perhaps drilling or mining operations. The polar regions of the earth are much easier to colonize than the moon, and interested groups might find solar power beamed from the sky a reason to build a colony at the South Pole. The military might use this as they have used the global positioning system (GPS): it is difficult to protect oil tanker trucks from harassment by even a minimally equipped group of determined attackers, while invisible power from the sky would be hard to obstruct.

In the larger picture of world energy supply, all of these analyses mean that space power satellites are not in any way economically competitive.

A similar conclusion was reached by Dr. Simon Peter Worden (http://www.thespaceshow.com/detail.asp?q=1127, http://www.nasa.gov/centers/ames/about/centerdirector.html.) (Brigadier General, USAF, retd.), Director of the NASA Ames Research Center (ARC), who stated on March 23, 2009, "Space based solar power is about 5 orders of magnitude more expensive than solar power in the Arizona desert." This is an estimate of $\$10^5$/W. Dr. Worden cited the high cost of putting materials into orbit.

6
Solar Cells Based on Single PN Junctions

6.1
Single-Junction Cells

The semiconductor properties described in Chapter 3 underlie the behavior of solar cells. Most relevant aspects are the formation and properties of PN junctions. In discussing leading types of single-junction solar cells, we start with Section 3.7, Equations 3.65–3.70, and Figures 3.16 and 3.17. Figure 3.17 shows the I–V curve under illumination of a silicon solar cell. We begin with silicon solar cells, which have evolved over time toward goals of improved efficiency and toward lower fabrication cost.

Figure 6.1 sketches the bands of a Si N^+P solar cell, built to be illuminated from the left, a cell that might be similar in its characteristic to that shown in Figure 3.17. The nomenclature and basic processes that go on in such a conventional single-crystal Si device were shown in Figure 3.16c. The device shown in Figure 6.1a in the dark, but Figure 6.1b shows the expected pattern of decaying intensity. It is important to understand that the absorption process strongly depends on the wavelength of the light, which is related to the photon energy through the relation $E = hc/\lambda$. The hatched regions on right and left indicate metallic contacts to connect to a load, and details are omitted, such as how the left contact is made optimally transparent to incoming light.

In practice, the front contact is applied as narrow fingers of metal, to allow light to enter. The front surface may be textured to reduce reflection and promote internal trapping of light as discussed below. In addition, the front surface may be coated with an antireflection coating (AR or ARC) typically with refractive index close to 2. Materials used on silicon include MgF_2, ZnS, Si_3N_4, Ta_2O_5, and TiO_2. The rear contact, here shown as a uniform metal layer, is sometimes replaced by an array of low-resistance contacts, with locally diffused P^+ regions. In this case, the larger portion of the rear surface will be oxidized to form SiO_2, quartz. The Si/SiO_2 interface reflects light back into the silicon and also, especially if passivated with hydrogen, reduces undesired recombination to a level less than would occur at a metal contact. The desired outcome for photogenerated electrons and holes is to be collected in the opposite terminals of the device and to drive an external current. Recombination

Nanophysics of Solar and Renewable Energy, First Edition. Edward L. Wolf.
© 2012 Wiley-VCH Verlag GmbH & Co. KGaA. Published 2012 by Wiley-VCH Verlag GmbH & Co. KGaA.

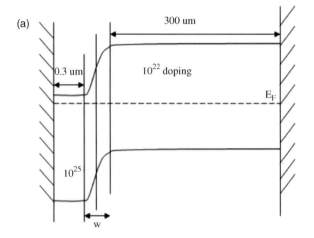

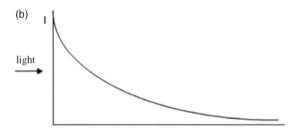

Figure 6.1 (a) Sketch of N⁺P junction solar cell, designed for illumination from the left side. The doping and dimensions are typical for a Si cell, see Figure 3.17. (b) Light absorption into the cell follows law $I = I_o \exp(-\alpha x)$, where decay constant α strongly depends on photon energy. At stated N⁺ doping 10^{25} m⁻³, the Fermi level in the N⁺ region actually will lie in the conduction band, making the overall band bending larger. (Courtesy of M. Medikonda).

internal to the device defeats this desired outcome. Recombination of photogenerated minority carriers is undesirable, and *surface recombination* can be inhibited by provision of a back surface field (BSF). In the case of the rear contact to the P-region shown in Figure 6.1, this could be provided by doping acceptor impurities to produce a P⁺ P junction at the rear. This would have the effect of raising the conduction band edge at the rear of the cell and thus reflect minority electrons back into the bulk, reducing the chance for surface recombination.

It is important to understand that the active region of the classic PN junction device extends well beyond the depletion layer W, to include the hole diffusion length L_p on the left, and the (minority carrier) electron diffusion length L_n on the right. The action of light absorption is to greatly increase the minority carrier concentrations (p_n on the left, n_p on the right) leading to an enhanced reverse current density (Equations 3.65–3.67). The photoelectric effect is essentially that a photon of light disappears and gives its energy entirely to an electron, and in the semiconductor context to create an electron–hole pair.

The radiation-induced current density of a solar cell is $J_L = -Ge(W + L_n + L_p)$, where L is the diffusion length for the minority carrier and W is the junction width. Looking at Figure 6.1a, when under illumination, G will represent the rate at which electron–hole pairs are created by light photons of energy exceeding the local bandgap energy. To be useful, the carriers have to be generated within a diffusion length L of the junction, where $L = (D\tau_r)^{1/2}$. Here τ_r is explicitly the minority carrier lifetime against recombination. L can be rewritten as $L = (\tau_r \tau_s k_B T/m)^{1/2}$ making use of the relations $\mu = e\tau_s/m = eD/k_B T$. (The scattering lifetime can also be written as $\tau_s = \lambda/\langle v \rangle$, using the mean free path and thermal velocity.) A long minority carrier lifetime is achieved by minimizing defects, which would facilitate recombination of electrons and holes, with particular attention to recombination at surfaces, as mentioned above. The total current density is the sum of the light-induced density and the thermionic current:

$$J = J_0[\exp(eV/k_B T) - 1] - Ge(W + L_n + L_p). \tag{6.1}$$

The reverse current density J_0, as discussed earlier, is strongly temperature dependent.

Setting $V = 0$ (short-circuit condition), the square-bracket term is zero, so we recognize

$$J_{sc} = -Ge(W + L_n + L_p), \tag{6.2}$$

where W is the depletion region width.

The open-circuit voltage V_{oc} is obtained by setting $J = 0$ in Equation 6.1, so that $\exp(eV_{oc}/k_B T) = 1 - J_{sc}/J_0$, and

$$V_{oc} = (k_B T/e) \ln(1 + J_{sc}/J_0). \tag{6.3}$$

From a basic point of view, this value cannot exceed E_g/e.

The output power per unit area is JV, and this is maximized at maximum power voltage V_{mp} (adjusted by the load), satisfying $\partial(JV)/\partial V = 0$, which gives

$$V_{mp} = V_{oc} - (k_B T/e) \ln(1 + eV_{oc}/k_B T), \tag{6.4}$$

and a corresponding current density J_{mp}, slightly reduced from the short-circuit value.

The ratio between the available power at optimum load and the limiting power is defined as the "filling factor" $FF = J_{mp} V_{mp}/J_{sc} V_{oc}$. In Figure 3.17, the FF is simply the ratio of the areas of the inner hatched region defining the optimum power point and the larger rectangle, defined by the short-circuit current density and the open-circuit voltage. The basic efficiency is

$$\eta = J_{sc} V_{oc} FF/P_{inc}, \tag{6.5}$$

where P is the incident light power, and

$$\text{FF} = [eV_{oc}/k_B T - \ln(1 + eV_{oc}/k_B T)]/(1 + eV_{oc}/k_B T). \tag{6.6}$$

Looking at these expressions, one can make a few observations on the efficiency. One must maximize the short-circuit current, which requires optimal absorption of photons and minimal recombination of the minority carriers before exiting the junction. One seeks a large open-circuit voltage $V_{oc} = (k_B T/e) \ln (1 + J_{sc}/J_o)$, which requires, in addition to maximizing J_{sc}, *minimizing the reverse current*

$$J_{rev} = J_o = e\,[n_p (D_n/\tau_n)^{1/2} + p_n (D_p/\tau_p)^{1/2}]. \tag{3.66}$$

Small reverse current density is promoted by making the thermal *minority* carrier concentrations small, favored by a low temperature (this is aided by making the *majority* dopings $N_{D,A}$ large) and also by a long recombination lifetime. In practice, the open-circuit voltage is limited by the built-in voltage V_B, because beyond that voltage the device no longer provides an exponential $I(V)$ relation in the forward bias regime. The upper limit of V_B is E_G/e, although, if both sides of the junction are doped into the metallic regime, the value could be slightly larger than this, given by Equation 3.59. To make an estimate of the optimal *filling factor* behavior, for 300 K and $E_G = 1.1$ eV for Si, we can take $eV_{oc}/k_B T = 42$. This gives FF $\approx [42 - \ln(43)]/43 = [42 - 3.76]/43 = 0.89$. One can see from this that large illumination, that is, concentration of the light using mirrors or lenses, is advantageous, to increase the open-circuit voltage and, hence, the conversion efficiency of the cell.

For the single-gap junction, the efficiency η approaches 30% for favorable assumptions, a result that dates to Shockley and Quiesser [63]. These authors also gave an "ultimate efficiency" of a more general nature, which is quoted as 44%, but is not as directly applicable to the single-junction solar cell as is the basic efficiency η (Figure 6.2).

These devices work by the photoelectric effect, the release of charge by annihilation of a quantum of light. Since light photons have nearly zero momentum, the absorption process favors semiconductors with a direct bandgap, where the electron and hole have the same momentum. Figure 6.3 shows the absorption coefficients for Ge, Si, and GaAs as a function of photon energy. GaAs has the typical direct bandgap behavior, desirable for a solar cell, while indirect bandgaps for Si and Ge lead to low optical absorption, especially at energies close to the bandgap value, 1.1 eV for Si. This means that the thickness of a silicon layer to completely absorb light is larger than that in direct bandgap semiconductors by the relatively low absorption constant. Faceting the surface to reduce reflection and to lengthen the path of the light within the cell is a means of overcoming this additional cost.

6.1.1
Silicon Crystalline Cells

The idea of surface texturing and its utility is illustrated in Figure 6.4. It is seen that the "facet" makes two reflections necessary for backscattering of light, so that if the

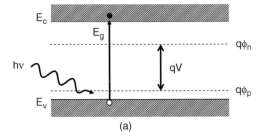

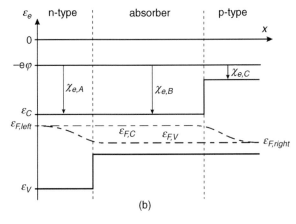

Figure 6.2 (a) The increased densities of electrons and holes under illumination can be described by separate "quasi-Fermi levels" φ_n and φ_p for electrons and holes, respectively. It is the shift of these separate quasi-Fermi levels that creates the output voltage of the cell, as indicated in (a). The separation of the quasi-Fermi levels arises as a competition between photogeneration and recombination, which may occur in the bulk or at surfaces. Recombination is to be avoided. Internal trapping and absorption of the light photons is to be maximized. (b) A generalized solar cell device, in which an interposed absorber layer creates electrons and holes from incoming light photons. Energy runs vertically, with zero at energy of free electron outside the device. This could be called a PIN device, with an (I) insulator layer. Work function φ shows highest energy of an electron in the device. Symbols χ are electron affinities in the different regions of the device, while E_C and E_V, respectively, are the lowest electron energy and the highest filled valence band energy. The effect of light is to separate the quasi-Fermi levels for electrons and holes, indicating nonthermal light-driven concentrations of electrons and holes. The resulting shift of quasi-Fermi energies gives the output voltage of the illuminated device [64].

single reflection occurs with 0.1 probability, the backscattering probability is reduced to 1%, a large improvement. The photon that enters the silicon now has a larger angle to the normal, making its path in the silicon longer before encountering the back surface, and also increasing its chance for internal reflection at that surface.

The reflection coefficient R is calculated at normal incidence by the formula $R = (n_1 - n_2)^2/(n_1 + n_2)^2$. Taking values 3.5 and 1 for the index of refraction for silicon and air, respectively, $R = 0.31$, which is substantial, and an antireflection coating is usually applied as mentioned above. The large index of refraction also implies that

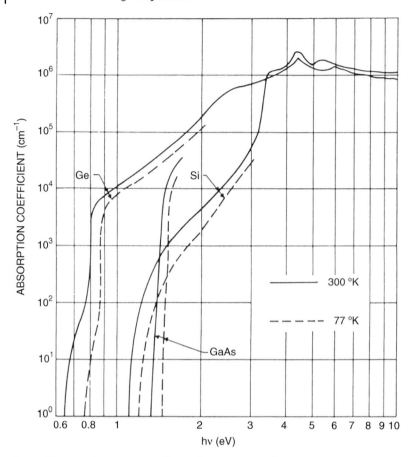

Figure 6.3 Optical absorption coefficients for Ge, GaAs, and Si at 300 K and at 77 K [65].

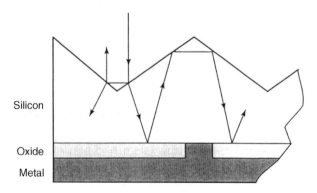

Figure 6.4 Surface-texturing stratagem for enhancing optical absorption and reducing external reflection of vertically arriving photons by faceting of light-admitting surface. This has been applied to (100)-oriented single crystal Si cells, and a similar method to polycrystalline Si cells, by M. Green and coworkers [66].

light inside the silicon is easily internally reflected. In fact, since the critical angle θ_c is given by $\sin\theta_c = 1/n_{Si}$ is small, $\theta_c = 16.6°$, all light except that within 16.6° of the surface normal is internally reflected. The corresponding angle at the Si/SiO$_2$ interface is 24.7°. Growing thermal oxide (quartz) on the back surface of the cell as indicated in Figure 6.4 reflects light to increase the light path in the silicon, and reduces the recombination rate of electrons and holes at the surface. Heating in a hydrogen atmosphere ("passivating") further reduces recombination at Si/SiO$_2$ interfaces, by filling "dangling bonds." Any step to increase the lifetime of minority carriers is beneficial to the operation of the solar cell, where the open-circuit voltage (the separation of electron and hole quasi-Fermi levels as in Figure 6.2) arises as a competition between photogeneration and recombination of carriers. (Figure 6.4 is simplified, neither the PN junction nor the front antireflection layer is shown, but these are clearly shown in Figure 6.5.) In Figure 6.4 an array of small contacts to the rear is seen.

The most efficient single-crystalline silicon solar cell, evolving from Figures 3.17, 6.1–6.3, is shown in Figure 6.5 [67]. The faceting indicated in Figure 6.3 has been implemented, showing the N$^+$P junction diffused into the faceted surface. The metal conductive fingers easily form ohmic contacts with the N$^+$ layer, and then the upper surface is coated with a MgF$_2$/ZnS antireflection double layer applied above a thin and

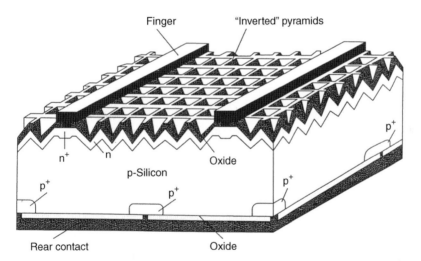

Figure 6.5 Structure of the best single-crystal Si solar cell, efficiency of 24.4%, developed by M.A. Green and coworkers. Note that entrance surface "oxide" is covered by an antireflection coating and that the larger part of the rear surface is oxidized to reduce recombination and also to reflect light back into the silicon wafer. The front top N-surface has locally diffused N$^+$ regions to facilitate formation of a low-resistance ohmic contact to metal fingers and also to collect current from the N-layer with minimal voltage drop and reduced recombination. On the top surface, the (thin-oxide plus antireflection layer) is shown as dark, while on the back surface thin thermal oxide (quartz) is indicated white and the rear metal contact is shown black. Locally diffused P$^+$ contacts connect the bulk P-region to the metallic back contact [68].

passivated native oxide (quartz) layer. This article also details a polycrystalline Si solar cell of 19.8% efficiency with a "honeycomb" texturing of the upper surface.

These authors found that completely enclosing the Si surface with thin thermal oxide to reduce recombination improved cell efficiency. However, the thermal oxide must be thin, on the order of 20 nm, so that the antireflection double layer applied above the oxide will still operate correctly. The polycrystalline version of this cell was grown on 1.5 Ω m, large-grained directionally solidified P-type silicon 260 μm in thickness. The use of diffused highly doped material just below metallic contacts, on front and back, suppresses recombination by repelling the minority carriers.

It was also reported that the Si/SiO_2 interfaces could be improved, passivated to reduce recombination of electrons and holes, by exposure to atomic hydrogen. Thus, the record-efficiency cell is denoted "PERL" (passivated emitter, rear locally diffused cell).

A large installation of single-crystal solar cells is shown in Figure 6.6, located at Nellis Air Force Base in the United States. This array provides 15 MW of power, and is shown to track the sun's motion in one direction.

The second example of a large solar panel installation, this time based on polycrystalline silicon (Dell Jones (2011) Regenesis Power, personal communication.), is suggested on the cover of this book, the lower right image. This view is similar to panels in a 2 MW field of 185 W Mitsubishi Electric modules, covering 16 acres, installed at the Florida Gulf Coast University in 2008. These modules are rated at 13.4% efficiency, and each consists of 50 cells of dimension 15.6 cm × 15.6 cm. These cells are thus similar to the silicon cells described above.

Figure 6.6 Nellis Air Force Base panels track the sun in one axis. Large conventional single-crystal silicon installation (http://en.wikipedia.org/wiki/File:Nellis_AFB_Solar_panels.jpg).

6.1.2
GaAs Epitaxially Grown Solar Cells

GaAs, as shown in Figure 6.3, has a suitable bandgap and direct absorption structure for solar cells. GaAlAs layers can be grown epitaxially by liquid and vapor-phase methods. Figure 6.7a shows an excellent cell using GaAlAs layers.

Features of this single-junction multilayer AlGaAs/GaAs solar cell include a thin p-AlGaAs window, grown by low-temperature liquid-phase epitaxy, "LPE," and a prismatic cover made of silicone (above the antireflection coating, "ARC"), which minimizes the optical losses caused by contact grid shadowing and reflection from the semiconductor surface. An efficiency of 24.6% was recorded with such cells, in a 100 times concentrated air mass zero, AM 0, spectrum. More details of the band structure in the cell are shown in Figure 6.7b.

6.1.3
Single-Junction Limiting Conversion Efficiency

Some crude estimates may be useful. For example, the open-circuit voltage has an upper limit E_g/e, where E_g is the semiconductor bandgap. The analysis [71] assumes a single junction and a single uniform bandgap energy. It is assumed that photons whose energy is less than the bandgap energy E_g do not contribute to the photocurrent. It is also assumed that the excess energy $hc/\lambda - E_g$ is lost to heat in the semiconductor. In practical terms, the junction structure should be thick enough that all of the light of energy $hc/\lambda - E_g > 0$ is absorbed.

A disadvantage of the single-junction cell at bandgap E_g is that all photons of energy less than E_g are lost, are not absorbed but pass through the cell. The energy output of the cell is the same for N photons of $2E_g$ or $3E_g$ as for N photons of energy E_g. For these reasons, the efficiency of the single-junction cell is inherently limited and depends on the spectrum of photon energies that are incident. Roughly, the bandgap for best efficiency, within the single-junction assumption, should be close to the peak of the incident spectrum.

Recalling Eq. 5.1 and Figure 5.2 the light spectrum from the sun is approximately that of a black body of temperature 5973 K, which peaks at $\lambda_m = (2.9 \times 10^6/5973)$ nm $= 486$ nm, which corresponds to an energy $1240/486 = 2.55$ eV. (The spectrum at ground level ("Air Mass 1") is substantially altered, by strong absorptions from several minor constituents of the atmosphere, including ozone, water vapor, and compounds of nitrogen and carbon. It is also weakened and redshifted by scattering in the atmosphere.) Adopting the hypothetical black body spectrum, for simplicity, if the bandgap of silicon is 1.12 eV, then the many photons in the range below 1.12 eV are completely lost, and an increasing fraction of the energy of the more energetic photons is also lost. The most probable photon at energy 2.55 eV will contribute not more than will a 1.12 eV photon, the difference energy 1.43 eV contributing only to heat.

The specialists in photovoltaic conversion have adopted an effective spectrum "Air Mass 1.5 G," correcting for an average daytime light path length 1.5 larger than when

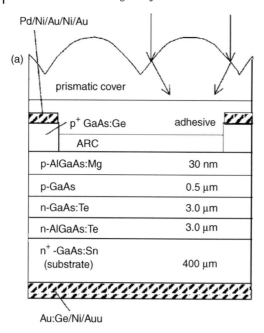

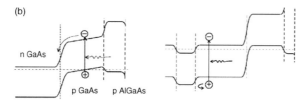

Figure 6.7 (a) Structure of epitaxially grown GaAs solar cell as utilized in spacecraft [69]. (b) Band diagram of epitaxially grown GaAs solar cell, efficiency of 24.6%. (*left*) The sloping band (built-in quasi-electric field) in the p-GaAs front layer, where photoabsorption occurs, was formed by Zn diffusion during LPE growth of the wide-bandgap AlGaAs window. (*Right*) Elaboration of cell with a back surface field provided by adding a highly doped N^+-GaAs layer. Light absorption in the rear N-layer is enhanced by the BSF, which inhibits recombination of minority holes with majority electrons at the rear surface. These diagrams make clear that light absorption primarily occurs in the field-free regions of dimensions L_p and L_n beyond the junction-forming depletion layer ([69], Figure 2.4, p. 25).

the sun is directly overhead. This spectrum is reduced from the top-of-the-atmosphere spectrum, Air Mass 0 (AM 0), by about 28%, of which 18% is from absorption and 10% from scattering. Scattering is well known to follow a λ^{-4} law, which removes blue light from the direct path but adds in some blue light scattered from the blue sky. The resulting AM 1.5 G spectrum (which also includes a diffuse light contribution) is used to calculate the ideal conversion efficiency of a single-junction solar cell operating at 300 K, as a function of the semiconductor bandgap energy. The results for the AM 1.5 G spectrum (labeled, in Figure 6.8) peaks near 29% efficiency at a

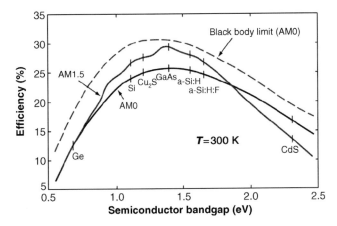

Figure 6.8 *Basic* conversion efficiency [70] of a single-junction solar cell at 300 K versus semiconductor bandgap energy E_G, due to Shockley and Quiesser originally. (Note that this plot is not the incident light spectrum! See Figure 5.2.) The dashed curve shows the basic efficiency curve if the input spectrum is taken as an ideal black body spectrum for $T = 6000$ K, a convenient theoretical assumption. AM 0 means sun spectrum in space, while "AM 1.5" is agreed test spectrum representing noon on a sunny day at a medium latitude, so the sun is not directly overhead at noon and the air traversal path length is about 1.5 the vertical height of the air mass. Dashed line represents best efficiency that could be obtained with fictitious black body sun spectrum 6000 K, while AM 0 is a measured spectrum above earth's atmosphere [70].

bandgap in the vicinity of 1.4 eV. This is close to the bandgap energy for GaAs, as is shown in Figure 6.8.

The efficiency of the single-junction cell, around 30% for one sun intensity, was addressed by Shockley and Quiesser [72].

The starting point is the diagram of the bands in the p–n junction, shown in Figure 3.16c. If we imagine the sketched open-circuit PN junction cell in thermal equilibrium at a temperature T, we realize that dynamical processes of thermal generation of electron–hole pairs occur throughout the structure to maintain the equilibrium distributions of electrons and holes in the various regions of the junction structure. The reverse current density J_{rev} comes from diffusion of minority carriers located within the diffusion lengths L of the junction, which carriers fall down the potential gradient. Again, electrons in a p-type semiconductor are of limited lifetime, since they can fall into an ionized acceptor site, give off light, and disappear. The statistics require a corresponding generation process to maintain the equilibrium concentration. The distance the minority electron in the p-region can diffuse before recombination is the minority carrier diffusion length L. Only carriers within such a distance of the actual junction can be usefully driven around the external circuit. At open circuit, a small forward bias appears, such that the net current across the junction is zero. The generation processes in equilibrium are balanced by recombination of electrons and holes to create photons. The recombination events can create photons of energy equal to or larger than $E_g = hc/\lambda$, where λ is the wavelength of the photon.

The insight of Shockley and Quiesser was to recognize that these internal processes must support emission of a black body spectrum, at least for wavelengths shorter than hc/E_G from this open-circuited thermal equilibrium structure. *The amount of light emitted, fixed by Planck's radiation law, is closely related to the reverse current density, J_{rev},* and the spectrum of the light, required by Planck's radiation law, will be related to the energy distributions of excited electrons and holes in the thermal equilibrium structure. If we recall Figure 5.1, and imagine the small black area A at the bottom of the figure to be the open-circuit solar cell at temperature T in the dark. It will act as a black body and radiate power outward, and the two cases of interest are with and without the concentrating optical system. In the absence of the optical system, a tiny fraction of the black body radiation will fall into an angular range comparable to that of the sun seen from earth, $f_s = 2.16 \times 10^{-5}$. With the optical system in place, all of the black body radiation will be focused into a solid angle comparable to that of the sun as seen from earth. According to Planck's law, the total power emitted by the junction of area A into the full hemispherical range of angles, solid angle $= 2\pi$, in the photon energy range E_1–E_2, is, writing power $P = dE/dt = E'$, based on Eq. 1.1 multiplied by $c/4$,

$$P(E_1, E_2) = E'(E_1, E_2, T) = A(2\pi/h^3 c^2) \int_{E1}^{E2} dE\, E^3/[e^{E/kT} - 1]\, \text{W}, \quad (6.7)$$

where $E_1 = E_G$. The upper limit on the energy may be approximated as large, $E_2 = \infty$, or may be estimated from the band structure of the semiconductor. The point is this completely determined number of watts W comes from band-to-band recombination in the solar cell, and gives a number that can be closely related [73] to J_{rev} (3.66)

$$AJ_{rev} = I_0 = eA(2\pi kT/h^3 c^2)[E_G^2 + 2kT\, E_G + 2(kT)^2] \exp(-E_G/kT). \quad (6.8)$$

It is also true that if the junction at temperature T acquires a voltage V between its terminals, then Equation 6.8 is simply multiplied by $\exp(eV/kT)$. We now write the current I in terms of the rate of change

$$dN/dt = N' = N'(E_1, E_2, T) = A(2\pi/h^3 c^2) \int_{E1}^{E2} dE\, E^2/[e^{E/kT-1}]\, s^{-1} \quad (6.9)$$

of photons exchanged between the area A junction at temperature T_c and the sun at temperature T_s, as in Figure 5.2.

$$I = Af_s\, N'(E_G, E_2 = \infty, T_s) - Af_c\, N'(E_G, E_2 = \infty, T_c)\, [\exp(eV/kT)^{-1}] \quad \text{Amperes.} \quad (6.10)$$

Set $f_s = 2.16 \times 10^{-5}$ to describe the absence of concentrating optics, and $f_c = 1$ to describe the area A radiating into all hemispherical directions 2π. The first term is the black body spectrum of the sun at T_s in the energy range above E_G. (See Equations 1.1 and 5.1 and related discussion.) The second term is the returning black body radiation of the cell at $T = T_c$ at its chosen open-circuit voltage V. We are describing an open-circuit cell, so $I = 0$. Taking the sun at 6000 K and the cell at 300 K, Shockley and Quiesser [72] thus found the best efficiency is 31% for $E_G = 1.3$ eV. The uppermost dashed curve in Figure 6.8 describes this result, using the vacuum given by Planck's law of solar spectrum.

Changing to the concentrating situation, where both f factors are 1.0 (input and output radiation both use a full hemispherical range of angles), the new equation is

$$I = Ae\, N'(E_G, E_2 = \infty, T_s) - Ae\, N'(E_G, E_2 = \infty, T_c)\, [\exp(eV/kT)^{-1}] \quad \text{Amperes.} \tag{6.10a}$$

It is now seen that a larger open-circuit voltage V is needed to balance the larger solar input from the concentrating optics. This means a larger efficiency, since the open-circuit voltage times the current is the power. In this case, Shockley and Quiesser thus find efficiency 40.8% at 1.1 eV for a fully concentrated single-junction solar cell; see Figure 5.1 for an example of such optics, where the imagined cell of area A is at the bottom of the field.

This analysis is based on the direct sunlight and does not include correction for the diffuse scattered light. This analysis is invalid for a cloudy sky.

In this way, the complex situation was analyzed to provide the maximum efficiency of the single-junction device as a function of the single bandgap energy, under assumed illumination conditions. The solid line plots of Figure 6.8 are numerically obtained from the Shockley–Quiesser analysis using the below-atmosphere spectra, as illustrated in Figure 5.2.

As we will see later, the most straightforward means of improving the solar cell efficiency is to put two or more single-gap cells in *tandem* (series connection), so that the highest energy photons are processed with the largest bandgap junction, and later cells process those photons whose energy was insufficient to generate electron–hole pairs in the prior junctions. A cascade of tandem cells can approach the efficiency of the Carnot machine, in principle. In practice, tandem cells of at least 40% efficiency under concentration have been demonstrated. The materials science and engineering of these tandem structures make them more expensive, but this can be counterbalanced by using concentrating light systems. The cells are more efficient at higher light intensity, because the open-circuit voltage increases with light intensity.

6.2
Thin-Film Solar Cells versus Crystalline Cells

An important class of solar cells are those made of thin films of semiconductors. These are cheaper because they use less of the primary material: a few micrometers rather than a fraction of a millimeter, but are generally of lower efficiency because recombination occurs at the grain boundaries. The grain size in polycrystalline films varies from millimeters to around a micrometer, but nanocrystalline films are also used with grains as small as 10 nm. Grain boundaries foster recombination of photogenerated electrons and holes, which occurs at defects such as dangling bonds at the surfaces. The dangling bonds can be mitigated by adding hydrogen to fill empty

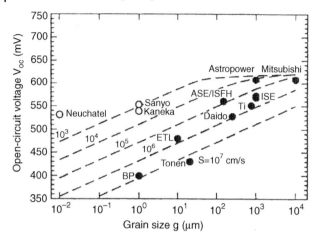

Figure 6.9 Empirical relation [73b] between open-circuit voltage and grain size for polycrystalline (closed circles) and nanocrystalline (open circles) silicon solar cells. Note that the grain size is as large as 10 mm for the Mitsubishi polycrystalline cells. Manufacturers and originating laboratories are indicated in the figure.

bonds, which allows the nanocrystalline films to be viable. Nanocrystalline films have the advantage that lower processing temperatures may suffice, making fabrication easier.

Recombination reduces output voltage attainable from a solar cell. Recombination is characterized by a velocity that may vary from 10^7 to 1000 cm/s, as shown in Figure 6.9 [73b].

Thin-film solar cells may also be made of organic compounds and polymers, as suggested in Figure 6.10b. Such cells are more easily made flexible and are potentially much cheaper. They have historically had low efficiency and, furthermore, usually have short lifetimes against degradation by the sun's rays. As we will see, several types of organic solar cells, including tandem cells, are well advanced in technology and have efficiencies approaching 10%. A basic problem with organics is that the electron and hole mobilities are low, so the materials are of low electrical conductivity. Sometimes, the polymers are loaded with conductive molecules such as C_{60} to improve the electrical conductivity. Such additions add to the cost.

Major opportunity for supplying competitive electric power from sunlight seems centered on thin-film devices, which are appropriate to reach a lower cost. (This conventional assumption may be challenged by the combination of large concentrating mirrors with extremely high-efficiency multiple junction tandem cells, as will be discussed later.) Thin films use much less material, and polycrystalline silicon cells are well known. (Polymer and dye-sensitized cells are inexpensive but seem to be limited in efficiency, as are amorphous silicon cells, at about 10%.) We here focus on perceived opportunities to reach a large market. We describe next a class of device that is capable of mass production.

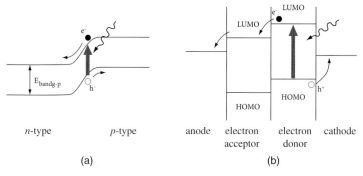

Figure 6.10 More general single-bandgap devices. Schematic [74] of p–n junction (a) and an organic bilayer structure (b) under illumination. Illumination leads to output voltage and current in external circuit. In (a), light can be usefully absorbed in wider region than the depletion layer, whose width includes the diffusion lengths for minority carriers in the semiconductor regions on both sides of the actual junction. Figure 6.10a applies to Si, GaAs, and $CuIn_{1-x}Ga_xSe_2$ "CIGS" cells, and to individual elements of tandem cells, generally of high efficiency. Figure 6.10b applies to "organic polymer" cells, generally of low efficiency and low cost. The third type of cell, the "dye-sensitized solar cell," is not shown. Multiple junction (tandem), "intermediate band," and quantum-dot-assisted cells are additional categories (see text).

6.3 CIGS ($CuIn_{1-x}Ga_xSe_2$) Thin-Film Solar Cells

Low-cost semiconductor cells with efficiency approaching 20% have evolved making use of alloys of selenides of copper, indium, and gallium ($CuIn_{1-x}Ga_xSe_2$, known as "CIGS"). These crystalline alloys are typically P-type semiconductors with appropriate bandgaps, high absorption coefficients, and quite large diffusion lengths. The cell is typically in the form of a thin N-layer placed near the front of an "absorber" of CIGS with thickness about 1 μm. The PN junction, as in Figure 6.10a, sends electrons and holes in opposite directions. The efficiency limit in these cells seems to be about 20%, with 15% available in preliminary large-scale production. This is in the same range as polycrystalline Si cells, but it appears that the CIGS cells are easier to fabricate on large scale and can be much cheaper to produce. These cells are made commercially by Avancis, Nanosolar, Honda Soltec, Solar Frontier, Wurth Solar, and Global Solar.

The historical origin of this line of CIGS cells may have been the CdS/Cu_2S cell formed by chemical surface treatment of CdS. This cell fell out of favor partly because of the toxic nature of Cd.

6.3.1
Printing Cells onto Large-Area Flexible Substrates

The reasons for ascendancy of CIGS include efficiency and durability, but the main reason is that these cells can be manufactured, without vacuum equipment, on a roll-to-roll mass basis similar to a modern printing press. The next figure shows one possible step-by-step prescription for $CuIn_{1-x}Ga_xSe_2$ CIGS cell manufacture that does not require vacuum equipment. (Many variations are possible, including those that avoid CdS and also those in which the substrate is simply aluminum foil.)

Figure 6.11, top to bottom, shows steps in one version of an ink-printing approach to CIGS solar cell formation. Substrates are either glass coated with Mo or a Mo foil, which can be thin and flexible. The "CIGS absorber layer" is formed in steps 2–5. The absorber is chemically $CuIn_{1-x}Ga_xSe_2$. The Cu, In, and Ga metal components are printed onto the cell in the form of oxide nanoparticles suspended in liquid, that is, a water-based ink. This is inherently efficient with regard to the use of the expensive elements, and the formulation of the ink allows close control of the composition. The layer is reduced (the oxygen is removed) by heating in a reducing atmosphere of

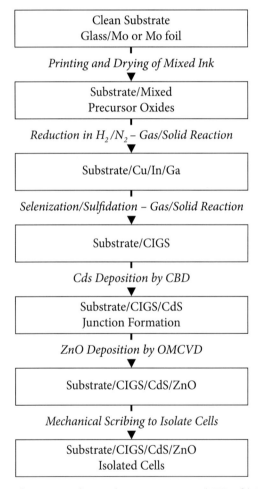

Figure 6.11 Ink-printed nonvacuum approach [75] to fabricate $CuIn_{1-x}Ga_xSe_2$ solar cells on flexible substrates.

nitrogen and hydrogen, so that oxygen leaves the film in the form of water vapor. The resulting metallic alloy layer is transformed to the semiconductor alloy by gas reaction at modest temperature with selenium using H_2Se gas, producing the "CIGS absorber." Speaking approximately, this is a P-type semiconductor of controlled bandgap and of thickness in the 1000 nm range. The thickness must be enough so that most of the light is absorbed. Again, speaking approximately, the device is a PN junction, and the next step, "Junction formation," is accomplished by deposition of CdS (CBD, or chemical bath deposition). The transparent electrode to the N-side of the junction is ZnO (OMCVD, organometallic chemical vapor deposition). The highest efficiency these workers achieved was 13.6% using a Mo-coated glass substrate. A variation of this process that does not use Cd is described next.

This basic process is important because it uses the minimum amount of the expensive metals, because it avoids expensive high vacuum equipment, and because it can be scaled up to large areas. Think of printing a newspaper, how many square meters of paper is printed by a major newspaper each day. The maximum efficiency reported for this type of cell is nearly 20%. This value is close to the value 20.3% mentioned for polycrystalline silicon cells, which are not amenable to a similar large-scale nonvacuum fabrication. It appears that some form of this basic ink printing process is employed in CIGS manufacture by Nanosolar, Inc. in California, and ISET, and for CdTe by a new firm, Solexant.

The cell shown in cross section in Figure 6.12 was formed on 12.5 μm thick commercial polyimide. The Mo back contact of 1 μm was deposited by DC sputtering. The CIGS layer was established following sequential evaporation of the metals Cu, In, and Ga, followed by evaporation of Se, with a controlled temperature anneal. The CdS layer was then deposited in a chemical bath process. RF sputtering was used to apply the upper window layer described as I-ZnO/ZnO:Al of 300 nm thickness. Finally, Ni–Al contact grids for better current collection were applied by electron

Figure 6.12 Scanning electron microscope cross-section image [76] of CIGS cell grown on polyimide flexible substrate. Efficiency of 14.1% was achieved in cells of this type. Layers, bottom to top: polyimide (not shown), Mo back contact, CIGS absorber layer, CdS junction-forming layer, ZnO insulator layer, and ZnO:Al conductive window layer.

beam evaporation. This is a complicated and expensive process but led to the most efficient CIGS cells, 14.1%, ever prepared on a flexible polyimide substrate.

The traditional process for forming the PN junction, the use of a chemical bath to deposit a layer of CdS, is objectionable because Cd is toxic. An alternative means of making the CIGS PN junction without Cd has also been demonstrated [77] achieving 16% efficiency. In this method, the CIGS absorber layer is contacted by a CIGS:Zn junction-forming layer, $Zn_{0.9}Mg_{0.1}O$ "insulator" layer, ITO (indium tin oxide) conductive window layer, and current collecting grid.

In this work [77], the CIGS absorber was deposited on Mo-coated glass by physical vapor deposition. The PN junction was formed by evaporating Zn onto the exposed surface of the CIGS, held at 300 °C. The zinc is believed to diffuse to a depth of about 50 nm in an annealing time about 5 min. This forms an internal PN homojunction within the CIGS. The $Zn_{0.9}Mg_{0.1}O$ "insulator" layer was grown on the CIGS:Zn surface by cosputtering the oxides from separate sources with adjustable rf power to establish the desired ratio $Zn/Mg = 9$. This ratio was found to adjust the conduction band position in the new layer to match that in the CIGS:Zn layer, to allow photoelectrons to pass out to the ITO electrode without scattering. The resulting physical interface was studied by transmission electron microscopy, TEM, and found to be epitaxial. The ITO transparent conductor about 100 nm thick was prepared by sputtering, followed by metal grid electrodes applied by evaporation.

High-efficiency CIGS cells are compared in Figure 6.14. (The same group has recently reported a record efficiency for this type of cell as 19.9%.) These cells use a coevaporation method for the CIGS layer. The traditional wet process is used to form the PN junction. This is described as growing 50–60 nm CdS films on the CIGS layer

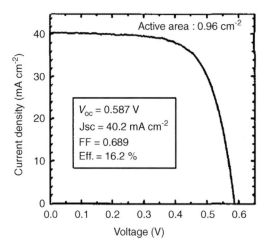

Figure 6.13 Current–voltage curve [78] of CIGS cell grown using a dry process and avoiding use of cadmium, on Mo-coated glass substrate. Efficiency of 16.2% was achieved in cells of this type after applying a MgF_2 antireflection coating. Layers, bottom to top: glass, Mo back contact, CIGS absorber layer, CIGS:Zn junction-forming layer, $Zn_{0.9}Mg_{0.1}O$ insulator layer, and ITO (indium tin oxide) conductive window layer.

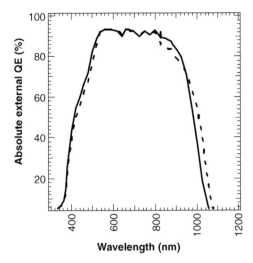

Figure 6.14 Quantum efficiency of CIGS solar cells [79]: solid curve, CIGS thickness 1 μm (17.2% efficiency); dashed curve, CIGS thickness 2.5 μm (18.7% efficiency). These cells benefit from a final antireflection coating of 100 nm MgF$_2$ after depositing the 200 nm thick window layer of Al-doped ZnO, which exhibited a sheet resistance 65–70 Ω/square, and a Ni/Al grid to collect current.

by immersion for 13 min in a 60 °C bath composed of 1.5 mM CdSO4, 1.5 M NH4OH, and 75 mM thiourea.

The authors report nuances in the method of coevaporation of the absorber to accomplish bandgap grading, and also suggest that for CIGS layer thickness below 1 μm the deep-level density increases. For this reason, mobility and lifetime are reduced when the thickness is reduced to 0.5 μm. This specific process is not scalable, but demonstrates a high efficiency, 19.9%, in single-junction CIGS solar cells.

The European firm Avancis has produced 15.8% efficient CIGS modules using a more conventional linear process. It appears that the firms using the ink process have not won customers, and, in fact, the total production of the CIGS cells is small compared to the production of CdTe cells, which presently dominate the thin-film solar cell market.

6.4
CdTe Thin-Film Cells

The largest thin-film solar cell supplier is First Solar, manufacturing thin-film cells of CdTe. According to the *New York Times* [80], First Solar, an American firm based in Tempe, Arizona, has signed an agreement with the Chinese government for a 2 GW photovoltaic farm to be built in the Mongolian desert.

The photovoltaic farm of area 25 square miles is part of a 11.9 GW renewable energy park to be built at Ordos City in Inner Mongolia. The overall project is to include 6.95 GW of wind power, 3.9 GW of photovoltaic power, and 0.72 GW of solar thermal farms. Further to the plan for Ordos City are biomass operations, fueled by organic materials like wood chips and straw for 0.31 GW, and 70 MW from hydro storage, a load-balancing technology that uses off-peak power to pump water to a high reservoir from which it can be released to turn turbines at peak demand periods.

First Solar is likely to build a plant in China to make thin-film solar panels. According to the article, the 2 GW solar farm as built in China is likely to cost significantly less than the $5–$6 billion if it were built in the United States. It is commented that the CdTe solar cells of First Solar are less efficient than the standard crystalline silicon solar cells made by companies like Suntech, but they are significantly less expensive to build. First Solar has also recently agreed to provide 1.1 GW to two California utilities from three big solar farms. It is commented in this article that the Chinese project is atypical of large solar projects, which have generally been awarded to solar thermal technology, which deploys mirrors to heat a liquid to create steam that drives an electricity-generating turbine, rather than to straight photovoltaic projects. It is pointed out in this comparison that the straight photovoltaic projects generally have fewer environmental impacts, such as requiring cooling water, and can be brought online faster than solar thermal plants.

Basic forms of CdTe cell construction are shown in Figure 6.15. The junction is between n-type CdS and p-type CdTe.

CdTe is a direct bandgap material with high absorption of light of energy beyond the bandgap about 1.45 eV. Although Cd is toxic, the devices contain thickness of only a few micrometers of CdTe and the toxic material is encapsulated to as not to present an external hazard. The material is very favorable it seems from the point of view of cost of fabrication. It appears that CdTe cells are leaders in the competition for large and cost-competitive photovoltaic installations.

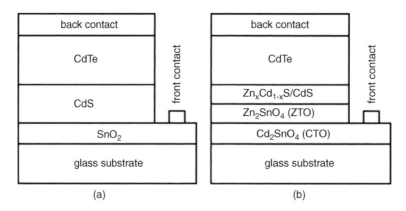

Figure 6.15 Two types of CdTe solar cells [81]. In both cases, the structure is built starting from the glass substrate, and methods including CSS (closed space sublimation) and chemical bath deposition are used to deposit succeeding layers. In these devices, the CdTe film is 1.5–3 μm in thickness, is deposited using CSS especially for large-area devices, and is subject to an anneal with $CdCl_2$ or other Cl- containing compound to promote grain growth after fabrication. Device efficiencies are in the range 10–16%, panel record efficiency is 12.8%.

6.5
Dye-Sensitized Solar Cells

Another potentially inexpensive form of solar cell is the dye-sensitized solar cell. This approach is based on an absorber that is partly titanium oxide, TiO_2 in its anatase tetragonal form, and partly dye molecules. Anatase can be prepared in a nanoporous form (sometimes, described as mesoporous, although microscopy shows particles of 10–80 nm diameter) by various procedures, one described as hydrothermal processing of a TiO_2 colloid.

This oxide has a bandgap of 3.2 eV, which means that the maximum light wavelength absorbed is $(1240\,eV\,nm)/(3.2\,eV) = 388$ nm. Thus, only a small part at the UV end of the solar spectrum is absorbed. Spectacular extension of the light absorption range to at least 700 nm has been demonstrated by coating (sensitizing) the nanoporous anatase with dyes. The dyes have a relatively low specific absorption, so that a thick layer of dye is needed, each dye being in intimate contact with the anatase surface so that a photoelectron can reliably be transferred to the conducting anatase.

An image is shown in the next figure of a nanoporous film, suitable for dye sensitization.

The hydrothermal deposition procedure for the film in Figure 6.16 started with hydrolysis of a titanium isopropoxide precursor and terminated with screen printing and firing of the semiconductor layer on a conductive transparent substrate. A conceptual schematic of a dye-sensitized solar cell based on nanoporous anatase is shown in Figure 6.17.

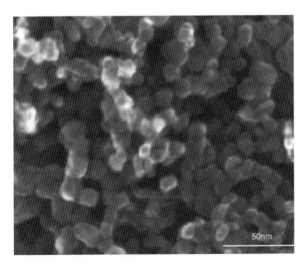

Figure 6.16 Scanning electron microscope image [82] of nanoporous anatase (tetragonal TiO_2) film prepared from a hydrothermally processed TiO_2 colloid. Note scale bar, which indicates particles about 10–80 nm in diameter. In dye-sensitized solar cells, dye must fully penetrate such a deposit, which may be 10 μm in depth, coating each anatase particle [82].

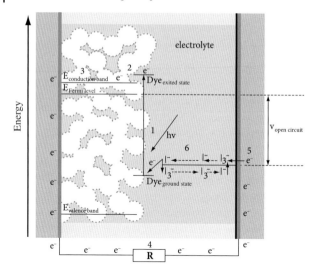

Figure 6.17 Principle of operation [83] of dye-sensitized solar cell. Note load resistor R at bottom, and light excitation of dye by photon $h\nu$ (step 1, center of figure). (*Left*) Schematic of nanoporous deposit of TiO_2 (titania, anatase form) coated with dye molecules (dots), and immersed in electrolyte containing iodine ions. Charge is carried to the dye by the iodide–triiodide reduction–oxidation couple in solution. Typically, the left side of the cell is commercial SnO_2-layered conductive glass, which has been coated with a Ti-nanoxide paste and fired [83].

6.5.1
Principle of Dye Sensitization to Extend Spectral Range to the Red

The operation of this device follows the steps indicated in Figure 6.17: step 1 is absorption of a photon by a dye molecule. Typically, the dye is a metal-organic ruthenium complex with response in the wavelength range 700–900 nm, light that is not absorbed by the titania. This is the idea of dye sensitization, to extend the absorption range of the photocell to utilize more of the solar spectrum. In step 2, the excited state electron jumps from the dye molecule into the conduction band of the TiO_2. The transfer is rapid from the dye only if the dye level lies ~0.2 eV above the conduction band, and reverse transfers from conduction band to dye are found to be slow. In step 3 the electron diffuses through the porous nanoparticle layer, which may extend 10 µm, and enters the electrode, and in step 4 the electron flows through the external load resistor R. In step 5, the electron is transferred into the electrolyte, which is typically aided by a very thin platinum catalyst layer. In step 6 the dye is regenerated into its ground state. From the diagram, it appears that the open-circuit voltage of the cell is approximately the photon energy divided by the electron charge, minus a voltage drop that occurs in the last step (step 6) when the redox couple brings charge back to the dye molecule. This may be oversimplified. Other factors including a potential difference across the back contact and a Fermi level shift of the anatase nanoparticles are discussed in an excellent review by Graetzel [82]. It is clear that these devices cannot be modeled in the conventional treatment outlined for crystalline semiconductor PN junction cells, where the diffusion lengths can be many µm, as outlined at the end of Chapter 3. It seems that low mobility and short diffusion lengths must necessarily limit the efficiency of dye-sensitized and other organic types of solar cells.

6.5.2
Questions of Efficiency

State-of-the-art performance of such a dye-sensitized cell is shown in Figure 6.18.

The dye-sensitized solar cell represented in Figure 6.18, under simulated AM 1.5 G irradiation, exhibited short-circuit photocurrent 20.5 mA/cm^2, open-circuit voltage 720 mV, and efficiency 10.4%. This efficiency is close to the best value ever obtained in this kind of cell. The dyes represented in Figure 6.18 are termed "black dyes" because of their wide absorption range, and are complexes based on ruthenium. The structure of a slightly different dye, N945, is shown in Figure 6.19.

Cells using this dye have been characterized by short-circuit current 18.8 mA/cm^2, open-circuit voltage 783.2 mV, and efficiency 10.8%.

6.6
Polymer Organic Solar Cells

Probably, the easiest type of cell to produce is the polymer organic cell, which involves coating a conductive substrate with two polymers and covering this with a transparent electrode.

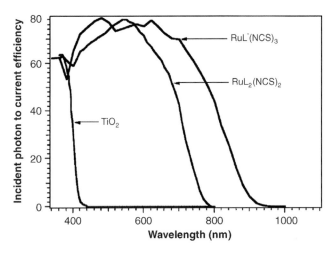

Figure 6.18 Principle of dye sensitization. Plots [84] of *incident photon to current efficiency*, for pure TiO$_2$ (left trace) and titania (anatase) coated with two different dyes (right two traces). It is evident that photoelectrons generated in the dyes are efficiently transferred to the titania, to flow in the external circuit. Dyes that are optimum have high absorption that extends to long wavelength, and many of the best performing dyes are based on the metal ruthenium. (Although the conversion of photons to electrons is efficient, these devices as solar cells exhibit relatively low efficiencies.) [84].

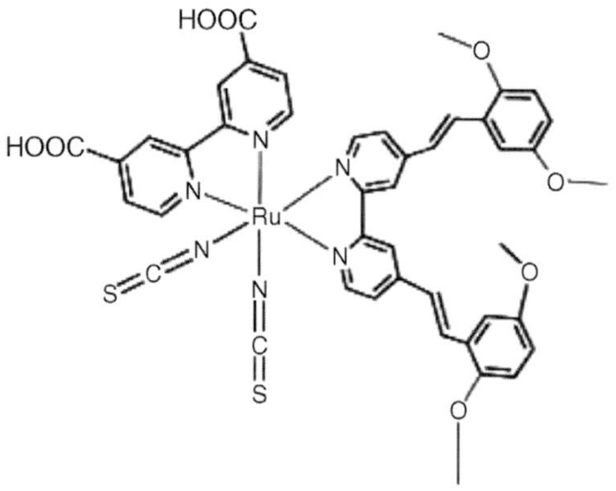

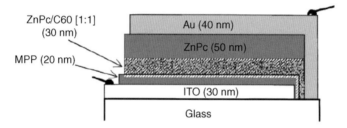

Figure 6.19 Ruthenium-based dye N945 [85], which is one of the best sensitizing dyes for TiO$_2$. This dye has a very strong absorption (a high extinction coefficient) allowing a more compact cell.

Figure 6.20 Schematic [86] of organic solar cell structure. This is a cell of the type shown in Figure 6.10b. Here, ZnPc refers to zinc phthalocyanine and MPP refers to a methyl-substituted perylene pigment. Both are available from major suppliers, and were purified by sublimation. The ZnPc/C$_{60}$ [1:1] layer was deposited by coevaporation of ZnPc and C$_{60}$. The ITO-coated glass substrate was obtained from a major commercial supplier. As described, this is not a low-cost process because of the vacuum equipment needed [86].

6.6.1
A Basic Semiconducting Polymer Solar Cell

The "ZnPc/MPP" organic solar cell depicted in Figure 6.20 exhibited an efficiency of 1.05% when illuminated with AM 1.5 radiation at an intensity of 860 W/m^2. The measured short-circuit current density was 5.2 mA/cm^2. This is an early example of a polymer organic cell. This type of cell was earlier shown in Figure 6.10b.

We defer further discussion of polymer organic solar cells to the next chapter, because the cell to be discussed is a good example of a tandem solar cell, an approach to reaching higher efficiency by making use of more than one bandgap.

7
Multijunction and Energy Concentrating Solar Cells

The single-junction solar cell wastes a bulk of the energy in the incoming light, as we have seen in detail in Chapter 6. Photons of energy greater than the bandgap lose the excess energy to heat, and photons of energy less than the bandgap go through the cell without absorption.

An intuitive solution to the problem is the opposite limit of a *series of cells with graded bandgaps*, so that there is in the stack somewhere a semiconductor that will optimally convert the energy of each photon. By extension of the Shockley–Quiesser analysis, this picture can be verified. In the limit of an infinitely subdivided tandem cell, that is a series connection of an infinite array of single-junction cells, the efficiency approaches, but does not reach, that of the Carnot cycle $\eta = 1 - T_c/T_s = 1 - 300/6000 = 0.95$. A concentrated infinite tandem cell is predicted in Figure 7.1 [87] to reach 86.5%, close to the 85.4% efficiency, called the Landsberg limit in connection with thermal solar power installations in Chapter 5. The curves in Figure 7.1 are numerically based on an AM -1.5 spectrum as shown in Figure 5.2. In the case of tandem cells, the calculation assumes that the power from each cell is extracted independently, while in practice the series-connected stack of cells is a two-terminal device and a constraint of equal current through each cell limits the power extracted. A small error may come from this difference, but in fact series-connected three-gap tandem cells have been demonstrated to operate at efficiency as high as 41.6% [88].

The record setting cell is a three-gap GaInP/GaInAs/Ge lattice-matched cell produced by Spectrolab, a subsidiary of Boeing Corporation, operated at 364-sun concentration. This is a structure grown epitaxially on a single crystal of Ge. We will discuss such cells later in this chapter. The junction technology is called liquid-phase organometallic epitaxy, and is an offspring of the general method of molecular beam epitaxy (MBE), used with success for compounds based on GaAs. It is credible that the efficiency in concentrating tandem cells with up to five different bandgaps may reach 50%. What is open to question is the market price that would apply to large-scale, complex systems of this type. It may also be mentioned that apart from the practical tandem arrangement of multijunctions, light-splitting optics can be used to illuminate a parallel array of junction devices. An efficiency record has been established by such an arrangement, although the cost associated is likely to be high.

Nanophysics of Solar and Renewable Energy, First Edition. Edward L. Wolf.
© 2012 Wiley-VCH Verlag GmbH & Co. KGaA. Published 2012 by Wiley-VCH Verlag GmbH & Co. KGaA.

7 Multijunction and Energy Concentrating Solar Cells

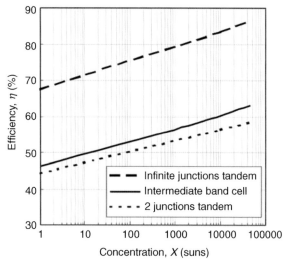

Figure 7.1 Efficiencies calculated for tandem cell combinations and the hypothetical "intermediate band cell." Note the common role of concentrated light intensity in improving the efficiency by raising the cell open-circuit voltage ([87], p. 121, Figure 6.7)

7.1
Tandem Cells, Premium and Low Cost

The most successful tandem cells have been those in the GaAs system, similar to the one mentioned as having achieved 41.6% efficiency at a concentration of 364 suns. Such a device would be used in a parabolic dish mirror concentrator, with an appearance similar to that of Figure 5.5. Tracking of the sun and water-cooling are implied.

7.1.1
GaAs-based Tandem Single-Crystal Cells, a Near Text-Book Example

The concept of the tandem cell is clearly shown in the spectral response curves published by the manufacturer Spectrolab, of the three-junction cell, shown in Figure 7.2.

The complex multijunction GaAs-based cells are at a high state of development, building on experience in the Soviet space program and in the wide application of III–V semiconductor (GaAs) materials in commercial electronics. See Figure 6.5 ff. in Chapter 6. An example close to the record 41.6% efficient cell is shown in Figure 7.3. The three main layers are evident from top to bottom, whose absorption spectra were shown in Figure 7.2. Complexities in this mature technology include the use of low-resistance PN junctions to series-connect adjacent cells, and provision for back surface fields to inhibit recombination. The problem of maintaining the three cells at optimum operating points consistent with identical current through each junction has been apparently addressed successfully.

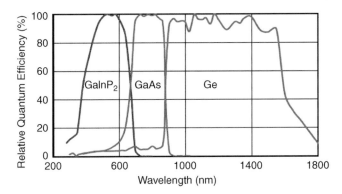

Figure 7.2 Relative quantum efficiencies (left to right) of the upper, middle, and lower light collecting layers of the GaAs-based three-junction tandem cell. This cell is epitaxially constructed such that a common atomic crystal lattice runs through the whole multilayer system, accomplished by organometallic liquid-phase epitaxy. The sum of the three spectra is seen to intersect most of the sun's spectrum. (Spectrolab press release. http://www.spectrolab.com/DataSheets/TerCel/tercell.pdf)

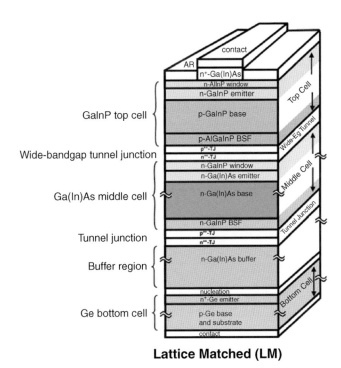

Figure 7.3 Structure of single-crystal three-gap GaInP/GaInAs/Ge lattice-matched cell. Cells close to this configuration have operated at 41.6 % efficiency at 364 sun illumination. The manufacturer Spectrolab is able to project four- and five-junction versions of such cells approaching 50% efficiency under concentrated illumination [89].

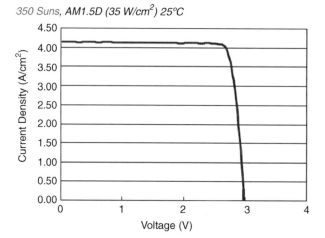

Figure 7.4 Current–voltage curve of three-junction GaAs cell under 350 terrestrial AM 1.5 sun illumination at 25 C. The open-circuit voltage, in principle, approaches $1/e$ times the sum of the bandgaps of the three junctions. (Spectrolab press release. http://www.spectrolab.com/DataSheets/TerCel/tercell.pdf)

Cells of the type shown in Figure 7.3 are termed "monolithic" and are two-terminal devices. It is evident that these complex devices are at a mature stage of development. What is not clear is what cost levels would be attainable with such devices in mass production. The complexity of the junction device is compounded, in an actual system, with the need for high-accuracy optics, suggested by Figure 5.1, and with a large and accurately tracking mirror or lens system suggested by Figure 5.8. Nevertheless, it is argued that with high concentration the actual cell area needed decreases to the point where the large installation could be cost-competitive with larger area simple flat plate cells such as those made by First Solar. The manufacturer Spectrolab says the tandem cells are useful both for point-focus concentrators and for dense arrays and linear concentrators (Figures 7.4–7.6).

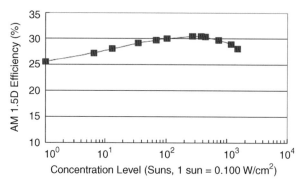

Figure 7.5 Measured efficiency as a function of illumination strength, for cell similar to that shown in Figure 7.3. A slightly different cell has achieved 41.6% efficiency as opposed to the 30% shown here. Note that the peak efficiency is achieved at illumination on the order of 300 suns. (Spectrolab press release. http://www.spectrolab.com/DataSheets/TerCel/tercell.pdf)

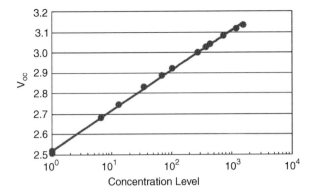

Figure 7.6 Measured open-circuit voltage V_{oc} as a function of illumination strength, for cell similar to that shown in Figure 7.3. A slightly different cell has achieved 41.6% efficiency as opposed to the 30% shown here. Note that these data support the enhanced power and efficiency in tandem cells predicted for concentrated illumination. (Spectrolab press release. http://www.spectrolab.com/DataSheets/TerCel/tercell.pdf)

It appears largely that the basic theory flowing from the work of Shockley and Quiesser has been validated by the results obtained from the GaAs-based tandem junction cells, as described.

What is not clear is what cost levels would be attainable with such devices in mass production. An attempt at cost analysis has been offered by the Spectrolab Boeing group [90].

The curves in Figure 7.7 are a claim by the manufacturer [90] that costs in a large-scale technology based on concentrating multijunction cells, with attendant tracking mirrors and water-cooling, can be competitive with large-scale installations such as

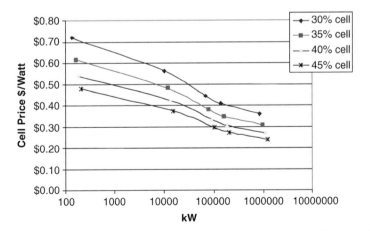

Figure 7.7 Projected cell costs [90] under different assumptions for efficiency (related to concentration) and volume of production.

the one being built by First Solar in China. The market has not supported this claim, and at present there appears to be no large-scale project of this type under construction.

7.1.2
A Smaller Scale Concentrator Technology Built on Multijunction Cells

A more modest system, in principle, similar to that described above is marketed by SolFocus. The individual parabolic glass mirrors are smaller and are packaged in arrays provided with two-axis tracking of the sun. A conceptual sketch for the SolFocus devices is shown in Figure 7.8. The literature suggests concentration in the range of 500–650. The dashed curve at the bottom of Figure 7.1 shows that a two-junction tandem cell at 500 suns has the capability of about 55% efficiency, a large improvement over single cells. SolFocus states the efficiency of their multijunction solar cells is 38%.

The SolFocus concentrating cell design addresses the problem of heating of the cell's PN junctions, by insertion of a *nonfocusing optical rod* to transmit the multiplied active spectrum, but not the unutilized and deleterious infrared heat spectrum, to the tandem junctions. Concentration of light (by about a factor of 500) with individual parabolic and accompanying secondary mirrors addresses the issue of cost and availability of elements, for example, Ga, used in the cells, and of the multijunction fabrication, since the concentration reduces the area of solar cell required. It appears that water-cooling is not required to operate the tracking panels provided by SolFocus because of the smaller scale and the use of the glass rod delivering the active spectrum from the mirrors to the cell.

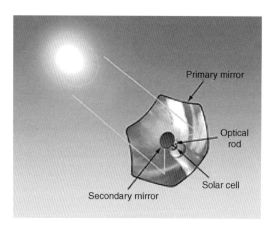

Figure 7.8 Artist's impression of concentrating solar cell including advanced GaAs-based tandem junctions. SolFocus markets arrays of these devices provided with two-axis tracking of the sun. There is a large change of scale between this approach and the approach represented by Figure 5.8. SolFocus, Inc., 510 Logue Avenue, Mountain View, California, 94043 (2008) http://www.solfocus.com/en/technology/.

According to the *Economist* [91], arrays of multijunction concentrating solar cells are available from Sunrgi, Inc., which are reportedly 37% efficient at 1600 sun concentration and can produce electricity at 5 c/kWh. (By comparison, the price of electricity in New York is 14 c/kWh.) It appears that the junctions are similar in design to those shown in Figure 7.3. As stated, this is a very favorable cost figure, which awaits confirmation.

7.1.3
Low-Cost Tandem Technology: Advanced Tandem Semiconducting Polymer Cells

The idea of a polymer solar cell was introduced in Figure 6.10, while a simple example of such a cell was shown in Figure 6.19. Such cells are usually considered among the least expensive to produce, and thereby are important in making solar energy more available. An improvement in this latter type of cell, to an efficiency of 6%, is our next topic [92].

The improved cell is of the tandem type, in which two cells are connected in series. Each of the constituent cells resembles that shown in Figure 6.10, but differs in that a network of internal heterojunctions pervades the charge-separating layers. The overall structure is similarly built on ITO-coated conductive glass and connected on the top with a thin metal electrode. The title of the article, "Efficient Tandem Polymer Solar Cells Fabricated by All-Solution Processing," makes a claim for efficiency and for a low-cost process.

The materials used in these cells are characterized as semiconducting polymers and fullerene derivatives. Typically, semiconducting polymers have a framework of alternating single and double C–C (sometimes C–N) bonds. Delocalization of the electrons in double bonds over the entire polymer molecule produces a molecular bandgap in the range of 1.5–3 eV. Light absorption in such a material produces an "exciton" or coupled electron–hole pair, which may diffuse but has to be dissociated in order to get a photocurrent. It is found that efficient exciton dissociation occurs if, instead of a single polymer, two separate polymers, called donor and acceptor polymers, are in contact. (This may be a well-defined interface, as in Figure 6.10 right panel, or a mixture of the polymers can create a network of such interfaces.) The exciton is generated in the donor material and charge separation occurs at the interface if the acceptor material has an empty energy level that is lower than the LUMO (lowest unoccupied level) of the donor (this is the level that is transiently occupied by a photoelectron). The exciton dissociation gives an electron in the acceptor material that is thus separated from the hole that remains in the donor. This charge separation gives an effective potential difference, induced by light, leading to a photovoltaic effect. In the work described [92], the acceptor polymers are loaded with fullerene molecules to improve the electrical conduction. The donor polymers, absorbing the light at different wavelengths by virtue of the chemical structure, act like dyes. The role of a dye was discussed in Figures 6.16–6.18. Exciton propagation is important in these materials, so that a dissociating interface can be reached, or else no external current will result.

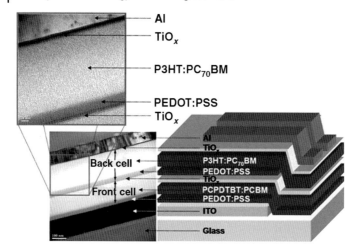

Figure 7.9 (Right) Schematic [92] of tandem polymer organic solar cell structure with efficiency over 6%. The structure consists of two cells (front and back) in series, separated by a transparent titanium oxide (TiO_x) layer. Each constituent cell is of the type shown in Figure 6.10, right panel, except that the charge-separating heterojunctions are distributed as a "bulk heterojunction composite" in the dye-fullerene layers (1st and 3D from top, in the schematic diagram). Here, PEDOT:PSS is a highly conductive hole transport layer and ITO is conductive indium tin oxide. The P3HT–fullerene and PCPDTBT–fullerene layers are the active charge separation layers, each described as a "bulk heterojunction composite" of a dye polymer donor with the fullerene acceptor. (Left, upper and lower) TEM cross-sectional images of portions of the structure. Note the sharp interfaces, and scale bars 20 nm (upper) and 100 nm (lower).

Light enters through the glass–ITO substrate into the "front cell" at the bottom. Absorption and charge separation in the front cell (PCPDTBT–fullerene) is strong in the UV and in the IR, but most of the visible spectrum passes through. The "back cell" (P3HT–fullerene) absorbs and charge-separates the visible portion of the spectrum. Thus, the two junctions harvest complementary portions of the solar spectrum. Because the cells are in series, and the electrical current is constant through the cell, the short-circuit current of the tandem cell is limited to the smaller of the short-circuit currents of the constituent cells.

In each cell, referring to right panel of Figure 7.9, photoelectrons flow vertically upward through the fullerene network of each composite and into the electron-conductive TiO_x layers. For the "back cell" (top of figure), these electrons go through the aluminum film into the external load. For the "front cell" the photoelectrons annihilate with holes at the interface of the TiO_x and the upper, hole-conducting PEDOT:PSS layer, which is fed photoholes from the (upper) "back cell." The annihilation of electrons and holes at the TiO_x–PEDOT:PSS interface makes the current continuous through the device. (The TiO_x layer will not transmit holes from the "back cell" into the "front cell" because its valence band energy, −8 eV relative to vacuum, is too low for holes to enter.) Photoholes generated in the "front cell" (lower) flow down through the lower PEDOT:PSS hole-conducting layer and annihilate at the

PEDOT:PSS–ITO interface with electrons from the external circuit that flow in through the ITO layer.

7.1.3.1 Band-Edge Energies in the Multilayer Tandem Semiconductor Polymer Structure

Some of this is clarified in an energy-level diagram for this complicated eight-layer structure, shown in Figure 7.10.

The absorption energy in the semiconducting donor polymer PCPDTBT (front cell, 3D from the left in Figure 7.10) is seen to be $4.9\,\text{eV} - 3.5\,\text{eV} = 1.4\,\text{eV}$, corresponding to wavelength $\lambda = hc/E = 1240/1.4\,\text{nm} = 886\,\text{nm}$. This creates an exciton, which migrates to a heterojunction interface with the acceptor. At such an interface, the photoelectron (resulting from decay of the exciton, and residing at 3.5 eV below vacuum level), falls across a heterojunction into the fullerene acceptor (PCBM) component of the bulk distributed heterojunction layer (energy 4.3 eV down from vacuum), where it can move freely to the conduction band of the TiO_x (down 4.4 eV from the vacuum energy). This can raise the conduction band energy of the fullerene layer by as much as $4.3\,\text{eV} - 3.5\,\text{eV} = 0.8\,\text{eV}$, to contribute to the open-circuit photovoltage of the cell.

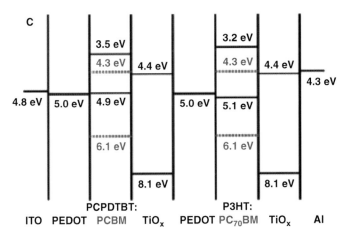

Figure 7.10 Energy-level schematic [92] of tandem polymer organic solar cell structure with efficiency over 6%. (Left to right in this diagram corresponds to rising vertically from bottom to top in Figure 7.9, right panel. So electrons flow from left to right in this diagram, and holes flow from right to left, under photoexcitation.) Energies are measured positively down from the vacuum level, so that 4.8 eV, on the left, is the work function of the ITO layer and 4.3 eV on the right is the work function of the aluminum film. For TiO_x layer (center) the valence band is at $-8.1\,\text{eV}$ (relative to free electrons in vacuum), which is too deep to allow holes flowing from the PEDOT layer to be transmitted, and these holes annihilate with photoelectrons from the front cell at the TiO_x–PEDOT interface, carrying current across the device. For the dye–fullerene "bulk distributed heterojunction" charge-separating layers (second from left, and second from right), the solid lines are the upper LUMO and lower HOMO energies of the donor polymer dye and the dashed lines are the conduction and valence band energies of the fullerene acceptors.

The photohole in the dye, residing at 4.9 eV below vacuum, has no chance of entering the valence band of the TiO_x (which would require an excitation of 8.1 eV − 4.9 eV), but it easily enters the hole-conducting PEDOT layer at 5eV. Similarly, in the "back junction" (3D from right) the absorption energy is 5.1 eV − 3.2 eV = 1.9 eV, corresponding to wavelength 1240/1.9 = 653 nm, closer to the visible. The subsequent movements of exciton, photoelectron, and photohole are the same as discussed for the "front junction." The maximum raising of energy across the back junction is 4.3 eV − 3.2 eV = 1.1 eV.

7.1.3.2 Performance of the Advanced Polymer Tandem Cell

The experimenters have measured the I–V characteristics of the tandem device, and also of each of the separate types of junctions, and find that the open-circuit voltage of the tandem device is close to the sum of the open-circuit voltages of the separate cells.

Tandem cell operating properties are clearly shown in Figure 7.11.

It appears that these efficiency values are the best ever obtained for polymeric organic cells of any type. The authors [92] discuss briefly that the tandem cell properties are reasonably stable over storage and running times, but note that any major application would require close attention to mechanisms of degradation.

7.1.4
Low-Cost Tandem Technology: Amorphous Silicon:H-Based Solar Cells

We have spoken earlier about the optical absorption of crystalline Si as being relatively small because of the indirect bandgap. This was shown in Figure 6.3 and led to the

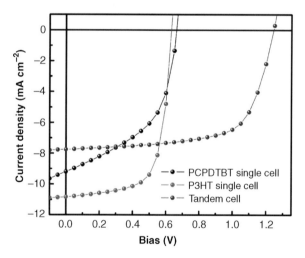

Figure 7.11 J–V characteristics [92] of single cells and tandem cell with PCPDTBT:PCBM and P3HT:PC_{70}BM composites under AM 1.5 G illumination from a calibrated solar simulator with irradiation intensity of 100 mW/cm^2 (about one sun). The tandem cell shows short-circuit current density 7.8 mA/cm^2, open-circuit voltage 1.2 V, and power efficiency 6.5%.

texturing of the surfaces to increase the distance a photon could travel inside the solar cell structure. Amorphous silicon can easily be deposited in thin layers and has a much higher optical absorption. This means that an amorphous silicon cell need not be as thick as a crystalline solar cell, a double advantage in reducing the cost of such a cell. The amorphous form of silicon, however, has inferior electrical properties, compared to crystalline silicon. Primarily, the carriers are less mobile because the material is not periodic in the location of the atoms. Related to the p–i–n nature of the junctions in these cells, one can refer to Figure 6.2a. The diffusion length described in Chapter 3 is short in the amorphous material, and a gradient of doping acting as a local electrical field is introduced to separate the photogenerated holes and electrons.

Several forms of amorphous silicon can be rather easily deposited that have characteristically different bandgaps. This facilitates making a tandem cell, based on the same principles as were described in connection with the GaAs triple-junction solar cells. Microcrystalline silicon, amorphous silicon, amorphous silicon passivated with hydrogen "a-Si:H", and Si:Ge all can be incorporated and have differing energy gap values. Amorphous silicon is slightly unstable, tending to the more stable crystalline state, and requires steps including hydrogen passivation, to fill dangling Si bonds in the incompletely filled bond set around the occasional atom, and also postfabrication annealing to allow a part of the reconstruction to occur before the device is put to use.

It should be said that these various forms of thin-film silicon solar cells are widely produced and employed in devices of all sorts. Where the efficiency is not a key parameter, useful products find many applications, from hand calculators to wristwatches to semitransparent glass panels that may be used in sky-lights or in a sun roof for a car. Also, especially connected with the United Solar Ovonics, LLC, company of Troy, Michigan, are inexpensive roofing tiles with built-in solar cells. The cells shown in Figure 7.12 are among the most efficient, approaching 11% efficiency.

The structures described here are wide-area devices produced on flexible stainless steel backing of 5 mil (0.127 mm) thickness and 35.6 cm width. This is a production process in which rolls of stainless steel foil are sequentially processed. Noting from Figure 7.12, the reflective coating Ag/ZnO is first applied, to reflect back into the cell light that has not been absorbed. The second full roll process was deposition of the three-layer silicon structure, as the foil advances down a line of processing stations including a radio frequency (rf) glow discharge to deposit Si at a thickness rate of 3A/s and a linear foil speed of 2.3'/min. This was followed by a full roll deposition of the transparent conductive oxide (TCO) to cover the structure. The roll was then cut into individual cells of dimension 35.6 cm × 23.6 cm, to which were applied wires and bus bars to collect current. The efficiencies measured on these large-area junctions were near 10.4% under one sun illumination, corresponding to open-circuit voltage 2.2 V and short-circuit current 5.7 A. (The best efficiency for cells of the type shown in Figure 7.13 is reported as 13% [94].) The large-area cells [93] are joined to form laminated modules for which the stable output power under one sun is expected to be 151 W. The larger open-circuit voltage, 2.2 V, coming from the series connection of junctions, enables a single cell to electrolyze water, which requires about 1.93 V, as

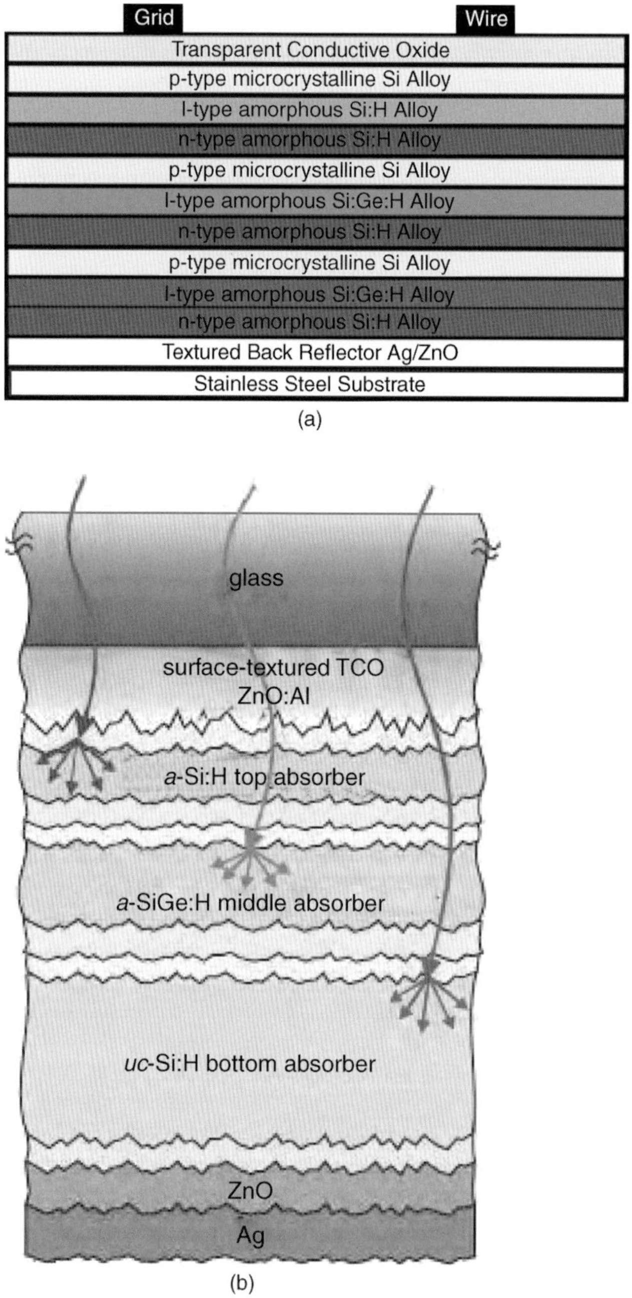

(a)

Grid		Wire
Transparent Conductive Oxide		
p-type microcrystalline Si Alloy		
I-type amorphous Si:H Alloy		
n-type amorphous Si:H Alloy		
p-type microcrystalline Si Alloy		
I-type amorphous Si:Ge:H Alloy		
n-type amorphous Si:H Alloy		
p-type microcrystalline Si Alloy		
I-type amorphous Si:Ge:H Alloy		
n-type amorphous Si:H Alloy		
Textured Back Reflector Ag/ZnO		
Stainless Steel Substrate		

(b)

glass

surface-textured TCO
ZnO:Al

a-Si:H top absorber

a-SiGe:H middle absorber

uc-Si:H bottom absorber

ZnO

Ag

Figure 7.12 (a) Structure of a tandem solar cell constructed with amorphous layers of Si and Si–Ge alloys [93]. Note that "insulator-type layers" are present in the structure, in which an effective electric field is maintained to counter the inherently low electron and hole mobilities in amorphous silicon and its alloys. (b) A similar tandem silicon cell including surface texturing to trap light [94].

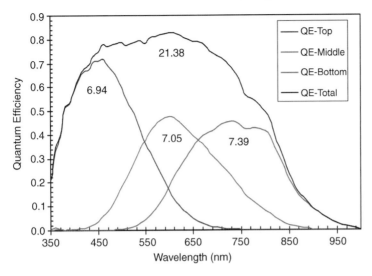

Figure 7.13 Quantum efficiency values measured in top, middle, and bottom of triple-junction silicon film cell structure of Figure 7.12 [93]. The different layers are deposited as the stainless steel foil substrate moves continuously through a line of deposition and processing stations, at a speed of about 2.3 feet/min.

described in Chapter 9. Single-junction silicon cells have an open-circuit voltage typically less than 0.6 V. Tandem silicon cells are used in the "artificial leaf" that we will describe in Chapter 9.

This kind of product could be used on the rooftops of New York City as was discussed in Chapter 5. This manufacturer, Ovonics, LLC, uses silicon, which is benign environmentally and of unlimited supply (sand is silicon dioxide), and this manufacturer also has a long history of providing solar tiling for roofs of apartment buildings. It appears that these flexible solar panels are available in 10 MW quantities. On the other hand, a lower price and higher efficiency $\sim$12.5% might well be obtained with thin-film single-junction CdTe cells, also available in large quantities.

7.2
Organic Molecules as Solar Concentrators

Here (Fig. 7.14) solar cells are mounted vertically on the edges of the light collecting plates, which have dissolved dyes. The upper plate collects blue photons, the dye reemits (fluoresces) and that light is sent to the cells on the edge. The fluorescence quantum efficiency of isolated dye molecules approaches unity. (Total internal reflection helps keep the reemitted light inside the plate, until absorbed in a solar cell).

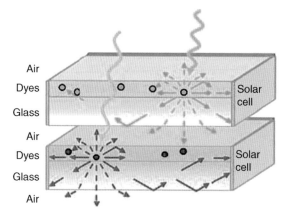

Figure 7.14 Concentration based on organic dyes that reemit light to solar cells at edges of glass light collecting layers [95].

The redder light passes through and is caught in the second plate that has a red dye. The effective collecting area is multiplied by this scheme, and the scheme also effectively involves two separate bandgaps, leading to higher efficiency in principle.

In the collection of the reemitted light from the dye molecules, two methods, fluorescence (prompt response) and phosphorescence (delayed response), are used. A dye molecule's characteristic color comes from the wavelength of its emission band. In the system shown in Figures 7.14 and 7.15, dyes absorbing blue and red, respectively, are located in upper and lower plates. The emission bands of these dyes match the absorption properties of the solar cells mounted around the edges of the light-collecting plates. A larger index of refraction in the glass promotes high efficiency in collection of the fluorescent and phosphorescent light by the solar cells, relative to its escape out of the glass into the vacuum. An effective area multiplication of around 50 is attained.

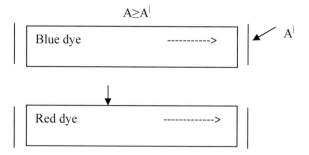

Figure 7.15 Area multiplication on the order of 50 by emission of fluorescent light and its capture in solar cells at the periphery [95]. (Sketch by M. Medikonda).

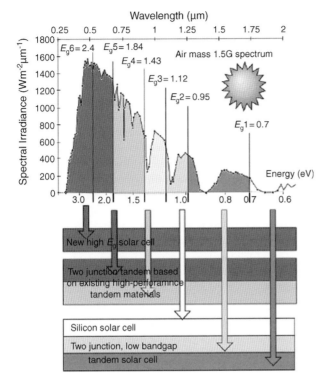

Figure 7.16 Schematic [96] of the multijunction solar cell grouping for the "lateral architecture." Upper panel shows the authors' version of the "AM 1.5 G" spectrum (see also Figure 5.2) with the authors' division of the spectrum into six ranges, indicating an optimum single-junction bandgap for each range. (Not shown, a semicylindrical dichroic lens system giving laterally displaced focal points for different wavelengths. The authors also mention arrays of these junction devices.)

7.3 Spectral Splitting Cells

A more elaborate, and surely more expensive, approach to multiple junctions has been described by Barnett *et al.* (see Figure 7.16) [96].

The new method involves splitting, by dispersive optics, the incoming light into "bins" of high energy, mid energy, and low energy. (In principle, the triangular glass prism of Isaac Newton would serve this purpose, and it appears that the efficiency of such optical devices can exceed 90%. In this case, a focusing action is also incorporated, to provide concentration of the light.) Light in each energy range is directed by such a "lateral optical system" to an appropriate, possibly multijunction, solar cell. In the schematic of Figure 7.16, two of the photon energy ranges are converted with tandem junction devices.

The same group reported [97] a record 42.8% efficiency for revised version of this "lateral optical system" approach, apparently involving five or six junctions in three separate laterally displaced devices, to which light is addressed by "spectral splitting optics."

7.4
Summary and Comments on Efficiency

In brief summary, the ideal single-junction cell at best captures only about 30% of the light energy. Going beyond the single-junction cell, a well-developed technology is to stack two or three junctions in series (tandem), engineered so that light not converted in the first junction may be converted in the second or third junction. This approach brings the measured efficiency for a (concentrated) tandem cell up to the vicinity of 41.8%. Tandem solar cells, typically based on GaAs, are commercially available but are expensive. On the other hand, in conjunction with mirrors or lenses to concentrate light, as mentioned in connection with Figure 7.7, this approach may possibly be competitive, even with electricity from coal-powered plants. Light dispersing optics has also been demonstrated to separate the spectrum and to steer different spectral regions to locations of appropriate solar cells.

The second, less developed, idea, is to use dye molecules in wide-area receiving plates, to funnel reemitted light to a dedicated array of solar cells. A large inventory of dye molecules has long been available, and this particular use of dyes is based on the most fundamental role of a dye, to convert "blue" or "green" photons to "red" photons. This proposal could turn out to be important.

The third approach, in principle, to improve efficiency is to multiply the number of charges produced per photon. This could make use of the blue portion of the solar spectrum, not fully utilized in the single-junction cell, on the conventional assumption of one charge per photon, with excitons. This effect is well known in the physics of the "Li-drifted germanium detector," a device for measuring the energy of a cosmic ray or other energetic particle by counting how many charges it generates by the exciton multiplication process in a pure sample of germanium. We will return to this topic in Chapter 8, and to the further possibility of an "intermediate band solar cell" (see the second curve in Figure 7.1).

7.5
A Niche Application of Concentrating Cells on Pontoons

A niche application of concentrating solar cells has been described [98] that exploits strengths and overcomes weaknesses of these devices. Solar cell efficiency decreases with temperature, which occurs in locations where the solar energy is strongest. Multijunction tandem devices definitely need cooling because of the concentration of energy, a factor of several hundred.

Concentrating devices need tracking that can be expensive. The pontoon-mounted concentrating solar cell arrangement shown in Figure 7.17 addresses and exploits these aspects. There are many locations where standing water is left open to the sun,

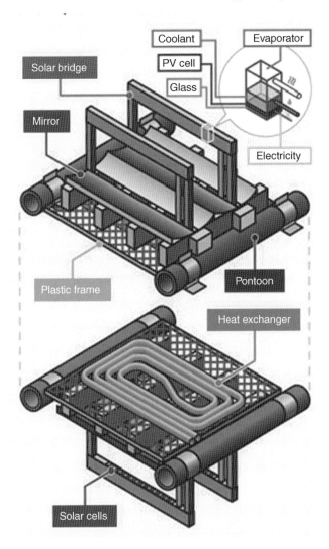

Figure 7.17 Schematic [98] of the multijunction solar cell application to pontoon mounting. Design of Solaris Synergy (Israel) to be built on a reservoir in France. Each unit is rated at 500 W, and circular groups of 24 units, held with hexagonal confining wires, are provided with sensor and motor to rotate as a unit. The arrays are rated at 12 kW, and the reservoir project contains eight arrays adding to 96 kW.

for example, at water reservoirs, above hydroelectric dams, in water treatment plants of various sorts, and in water transport systems. These areas, a bit like the top of a landfill, are often left open to the sun and can be exploited without undue complications. The water surface is reliably horizontal and tracking of the sun is accomplished by rotating the circular array of 24 floating pontoons around the vertical axis, which is easy to do. The device shown places the expensive solar cells in a bridge at the focus of parabolic mirrors, as shown, and water-cooling with heat exchange coils under the pontoon is a central feature. So, this device will work in hot regions where the sun is strongest but only on a water surface. This is a niche application, a small market that may be relatively easy to open to concentrating solar cells. Worldwide there are many such locations, often favorably controlled with respect to gaining access.

For example, the Colorado River aqueduct, part of the 400 mile California aqueduct, is a 242 mile water conveyance system that provides a large portion of the drinking water in southern California, diverting water from the Colorado River. About 63 miles of the Colorado River aqueduct is an open canal, on the order of 100 feet wide, too shallow for boating, and open to the sky. The solar power at 200 W/m^2 falling on this area of water is then on the order of 724 MW. This large area of water is under the control of a single authority that might be persuaded to make use of its potential as photovoltaic power source. An account of planning for exploiting this waterway for solar power has recently been given [99]. The company Solaris Synergy, mentioned in connection with Figure 7.17, estimates that the California aqueduct could yield 2 MW per mile, totaling up to 800 MW, and that the controlling authority is concerned that the floating solar arrays would need to be moored on the banks, which is the method used by Solaris Synergy to extract the electric power.

8
Third-Generation Concepts, Survey of Efficiency

What can be done to make solar cells more efficient? We have already described the main classes of solar cells that are now available. These can be characterized in terms of the number of distinct energy gaps, according to the physical form, and according to the use for single sun illumination versus concentrated illumination.

New conceptual approaches, which may become available in practice, may lead to higher efficiency and lower cost. These were suggested in Figure 5.6 in the category "advanced thin films." The agreed criterion for "third generation" designation is that the efficiency exceeds that for the single-junction cell, about 30%. At present, the only cells to achieve "third-generation" efficiency are, in fact, single-crystal tandem cells and spatially dispersed multijunction cells, at most ~41%. The tandem cell is a proven device, as we saw in Chapter 7, but at high demonstrated efficiency it has been available neither in thin-film form nor at low cost.

The first new concept that may alter this situation is "the intermediate band cell." The second is some form of carrier multiplication, as a means of harvesting in the external circuit the photon energy in excess of the bandgap.

8.1
Intermediate Band Cells

Efficiency comes from utilizing a larger portion of the solar spectrum. The idea of the intermediate band cell is suggested by the improvement experimentally realized in the two-bandgap tandem cell. The generalization is that some homogeneous semiconductor might be found to have three characteristic energies, E_C, E_V, and E_{IB}, so that the differences among these energies could represent two or three separate gaps, possibly acting to intercept a larger fraction of the solar spectrum.

A pioneering calculation along these lines is summarized in Figure 8.1, based on a hypothetical semiconductor as illustrated in Figure 8.2. (We will see in Section 8.3 that dilute magnetic semiconductors may be workable examples of such a band structure.)

The concept of the intermediate band semiconductor is shown in Figure 8.2. The Fermi level lies in the *intermediate band*, which is metallic, and is partially filled. In semiconductor physics, this situation arises in a semiconductor with a large

Nanophysics of Solar and Renewable Energy, First Edition. Edward L. Wolf.
© 2012 Wiley-VCH Verlag GmbH & Co. KGaA. Published 2012 by Wiley-VCH Verlag GmbH & Co. KGaA.

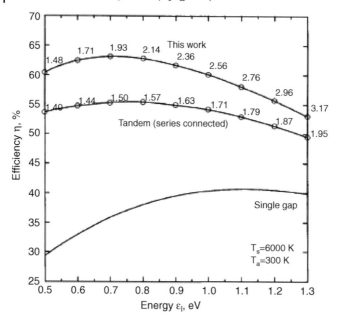

Figure 8.1 Efficiency of intermediate band solar cell [100] (top curve, labeled "This work") compared with two-junction tandem cell (second curve) and single-junction cell (lower curve) at one sun illumination. The abscissa is the value of the smaller gap, with the values for the second (larger; labeled E_H in Figure 8.2) gap in eV printed along the upper two curves, which assume two separate gaps. These curves are calculated assuming a black body spectrum at 6000 K [100].

concentration of donor impurities, such as P in Si. At a level of doping above the Mott transition concentration, the electronic states are linear combinations of electron waves centered on the impurity sites, and are delocalized. The Fermi level then lies within the distribution of impurity ground-state levels. In that case of large donor concentration, the energy gap between the Fermi level and the conduction band edge

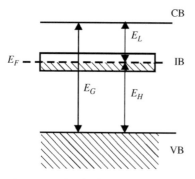

Figure 8.2 Sketch of hypothetical intermediate band semiconductor. Adapted from Ref. [101], Figure 7.1a, p. 141.

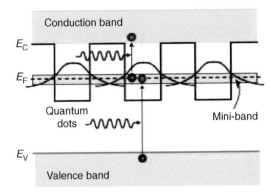

Figure 8.3 Two separate photon absorptions are illustrated in schematic diagram of intermediate band solar cell, as realized using coupled quantum dots to form a "mini-band." The idea is that the two smaller gaps allow utilization of the part of the fixed solar spectrum below $E_C - E_V = E_G$. In this device, the usual absorption for $hc/\lambda \geq E_G$ also occurs, at least in principle. The upper curve shown in Figure 8.1 assumes that carriers in the intermediate band have high mobility [102].

would be small, in the range of meV, as was discussed in Chapter 3 in relation to hydrogenic impurities. A similar situation might occur with chemical impurities that form deeper levels within the gap. In the work cited, the authors suggest the intermediate band could be formed by a regular array of quantum dots (QDs), three-dimensionally localizing electron states.

The small overlap between adjacent dots is assumed to form the intermediate band (referred to as mini-band) in Figure 8.3.

A sketch of an intermediate band quantum dot solar cell device is given in Figure 8.4. It is assumed that the metallic intermediate band results from half filling with electrons the quantum dot wells, as indicated. In this structure, the intermediate band is isolated from the external contacts by means of heavily doped p- and n-type emitter layers.

Experimental evidence of photocurrent from subbandgap light in an intermediate band solar cell structure has recently been demonstrated [104].

The authors [104], however, state that their devices do not exhibit the main features expected for the intermediate band solar cell. The results so far obtained, thus, give little indication that a practical intermediate band solar cell device is near. Since it is difficult to fabricate a complicated structure as asked for in Figures 8.3–8.5, the failure of the experiment may reflect the difficult fabrication rather than a failure of the essential intermediate band cell concept.

8.2
Impact Ionization and Carrier Multiplication

The "Li-drifted germanium detector" (http://en.wikipedia.org/wiki/Germanium_detector#Germanium_detector.) is an elegant commercially available solid-state

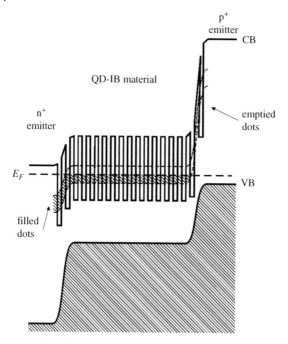

Figure 8.4 Sketch of intermediate band quantum dot solar cell device. Adapted from Marti et al. [103], Figure 7.12, p. 161.

device to measure the energy of a single high-energy gamma ray photon. A pure single crystal of germanium is treated in a special way with lithium, so that the mean free paths of electrons and holes are larger than the dimensions of the device. Metallic electrodes apply a small bias voltage across the germanium, and a sensitive detector of the charge flow in the circuit is provided. The germanium is pure and is also cooled to 77 K, the temperature of liquid nitrogen, so that the numbers of thermally generated free electrons and free holes is low, and only a small background current flows. A gamma ray, that is, a photon of high energy (MeV) $E_\gamma = hf \gg E_g$, is absorbed, creating initially an electron in the conduction band and a hole in the valence band, whose kinetic energies add up to $hf - E_g$. The final result from the single absorbed gamma ray photon, an outcome of internal processes in the germanium, is n electrons and n holes flowing in opposite direction across the crystal and into an external charge meter, such that the total measured charge

$$Q = n2e = 2e(hf/E_g) = \int I(t)dt. \tag{8.1}$$

From this relation, the gamma ray energy hf is deduced from the measured charge Q and the known value of E_g, according to

$$E_\gamma = (Q/2e)E_g. \tag{8.2}$$

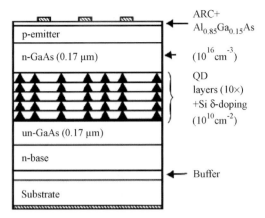

Figure 8.5 Example of intermediate band solar cell [104] based on quantum dots. Here, "ARC" represents "antireflection coating" and "δ-doping" means a pure deposit, in this case of Si. The indicated quantum dots are InAs, which have also been used to lower the threshold current in versions of GaAs heterojunction lasers. The authors [104] found extra photocurrent in this structure arising from lower energy photons, substantiating an aspect of the model depicted in Figures 8.2–8.4.

The carrier multiplication process that occurs within the extremely pure semiconductor creates $n = (hf/E_g)$ electron–hole pairs from the initial high-energy electron–hole pair. The high-energy carriers reduce their kinetic energy by collisions creating additional electron–hole pairs of lower energy, and in this process loss of energy to lattice vibrations (phonons) apparently is not an important effect.

A similar process would be beneficial in the efficiency of solar cells, to capture energy from the portion of the solar spectrum in the range $hf > E_g$. This process in a solar cell could multiply the conversion efficiency by a large factor.

The idea of a quantum dot was given in connection with Equations 2.11–2.16, and later in section 3.3.1 of Chapter 3. The wavefunction Equation 2.16 and energy level Equation 2.11 of a single electron in a confining potential of width L is given in Chapter 2 and generalized to three dimensions in Chapter 3, see Equation 3.15.

In the present context, quantum dots are 3D structures introduced locally in a junction device such as a solar cell or junction laser.

A quantum dot may be called an "artificial atom," characterized by discrete sharp electron energy states, and sharp absorption and emission wavelengths for photons.

Transmission electron microscope (TEM) images of such nanocrystals, which may contain only 50 000 atoms, reveal perfect crystals having the bulk crystal structure and bulk lattice constant. Quantitative analysis of the light emission process in QDs suggests that the bandgap, effective masses of electrons and holes, and other microscopic material properties are very close to their values in large crystals of the same material. The light emission comes from radiative recombination of an electron and a hole, created initially by the shorter wavelength illumination.

The energy E_R released in the recombination is given entirely to a photon (the quantum unit of light), according to the relation $E_R = h\nu = hc/\lambda$. Here, ν and λ are, respectively, the frequency and wavelength of the emitted light, c is the speed of light 3×10^8 m/s, and h is Planck's constant $h = 6.63 \times 10^{-34}$ J s $= 4.136 \times 10^{-15}$ eV s. The color of the emitted light is controlled by the choice of L, since $E_R = E_G + E_e + E_h$, where E_G is the semiconductor bandgap, and the electron and hole confinement energies, E_e and E_h, respectively, become larger with decreasing L as $h^2/8mL^2$.

These confinement (blueshift) energies are proportional to $1/L^2$. Since these terms increase the energy of the emitted photon, they act to shorten the wavelength of the light relative to that emitted by the bulk semiconductor, an effect referred to as the "blueshift" of light from the quantum dot. These nanocrystals are used in biological research as markers for particular kinds of cells, as observed under an optical microscope with background ultraviolet light (UV) illumination.

In these applications, the basic semiconductor QD crystal is coated with a thin layer to make it impervious to (and soluble in) the aqueous biological environment. Another coating may then be applied that allows the QD to preferentially bond to a specific biological cell or cell component of interest. The biological researcher may, for example, see the outer cell surface illuminated in green while the surface of the inner cell nucleus may be illuminated in red, all under illumination of the whole field with light of a single shorter wavelength.

8.2.1
Electrons and Holes in a 3D "Quantum Dot"

Equations 2.11–2.16 and 3.15 are applicable to the electron and hole states in semiconductor "quantum dots," which are used in biological research as color-coded fluorescent markers. Typical semiconductors for this application are CdSe and CdTe.

A "hole" (missing electron) in a full-energy band behaves very much like an electron, except that it has a positive charge, and tends to float to the top of the band. That is, the energy of the hole increases opposite to the energy of an electron.

The rules of quantum mechanics that have been developed so far are also applicable to holes in semiconductors. To create an electron–hole pair in a semiconductor requires an energy at least equal to the energy bandgap, E_g, of the semiconductor.

We found earlier that the wavefunction for a particle in a three-dimensional infinite trap of volume L^3 with impenetrable walls are given as

$$\psi_n(x, y, z) = (2/L)^{3/2} \sin(n_x \pi x/L) \sin(n_y \pi y/L) \sin(n_z \pi z/L),$$

where $n_x = 1, 2 \ldots$, etc.,

and

$$E_n = \left[h^2/8mL^2\right]\left(n_x^2 + n_y^2 + n_z^2\right).$$

8.2 Impact Ionization and Carrier Multiplication

The conceptual leap that is required here is that instead of a free electron in a vacuum box of side L with infinite potential walls, we have a conduction electron free to roam about in a cube of intrinsic semiconductor of side L, with the electron being confined to the conductor by the work function barrier. The moving carrier will be endowed with an effective mass by its immersion in the semiconductor, and it will be affected also by the permittivity of the semiconductor medium.

(Different wavefunction and energy equations will apply to other geometries than a cube, for example, the quantum dots grown by molecular beam epitaxy (MBE) are often pyramids. The wavefunctions in these cases will be different, but the differences in the end do not matter very much. Sharply defined energies inversely proportional to the square of the container size will result.)

This application to semiconductor quantum dots requires L in the range of 3–5 nm, the mass m must be interpreted as an effective mass m^*, which may be as small as $0.1 m_e$. The electron and hole particles are generated by light of energy

$$hc/\lambda = E_{n,\text{electron}} + E_{n,\text{hole}} + E_g. \tag{8.3}$$

Here, the first two terms are strongly dependent on particle size L, as L^{-2}, which allows the color of the light to be adjusted by adjusting the particle size. The bandgap energy E_g is the minimum energy to create an electron and a hole in a pure semiconductor. The electron and hole generated by light in a bulk semiconductor may form a bound state along the lines of the Bohr model, described above, called an exciton. However, as the size of the sample is reduced, the Bohr orbit becomes inappropriate and the states of the particle in the 3D trap are a more correct description.

In this context, Figure 8.6 shows levels in a quantum dot as an element of absorber in a solar cell. The process shown is one of absorption of a high-energy photon

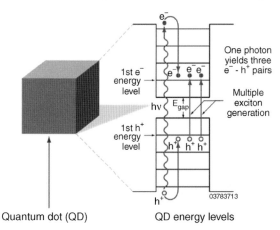

Figure 8.6 Multiple exciton generation in a quantum dot [105]. Because of quantum confinement, the energy levels for electrons and holes are discrete. A single absorbed photon of energy at least three times the energy difference between the first energy levels for electrons and holes in the quantum dot can create three excitons, tripling the charge in the external circuit.

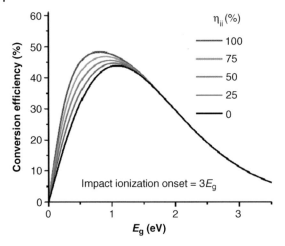

Figure 8.7 Predicted solar cell efficiency as enhanced by impact ionization. Evidence for such processes has been advanced for nanocrystals of the small bandgap semiconductor PbSe [106].

$hc/\lambda = E_g + 6\Delta$, where Δ is the energy spacing of succeeding size-quantized states in the quantum dot. The process depicted leads to three electrons compared to one electron eligible to flow through the load resistance, but requires an energy threshold $E \geq E_g + 8\Delta$, and the further condition

$$6\Delta \geq 2E_g + 4\Delta, \quad \text{or} \quad \Delta \geq E_g.$$

The lower threshold to produce two electron–hole pairs would be $E \geq E_g + 4\Delta$, with

$$4\Delta \geq E_g + 2\Delta \quad \text{or} \quad 2\Delta \geq E_g.$$

The minimum excess energy released by exciton decay to give charge multiplication is $E_g + 2\Delta$. This is a high threshold: since $2\Delta \geq E_g$, the initial energy has to be more than $3E_g$. The further assumption is that the charge appearing at the lowest electron and hole energies will efficiently leave the quantum dot to flow in the external circuit. This may be possible but apparently has not been demonstrated in situations other than the Ge detector.

If such processes are important, the optimum solar cell would have smaller values of E_G, as shown in Figures 8.7 and 8.8.

8.3
Ferromagnetic Materials for Solar Conversion

A recent theoretical investigation has concluded that dilute magnetic semiconductors may be useful as photovoltaic materials, having the possibility of the Fermi level located in a narrow magnetic band inside a larger gap. The compound AlP:Cr has been investigated, even though it has not apparently been synthesized. The main features of the band structure are shown in Figure 8.9 [107]. The electron band

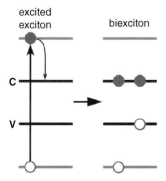

Figure 8.8 Basic process of impact ionization, as observed in PbSe nanocrystals [106]. This process is also involved in the germanium detector mentioned earlier.

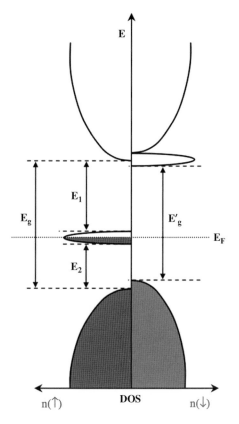

Figure 8.9 Sketch of electron density of states for ferromagnetic dilute magnetic semiconductor AlP:Cr as suitable for a solar cell [107]. AlP is a member of the GaAs family of semiconductors and is capable of being doped with Cr to become ferromagnetic. The narrow level in the bandgap is a d-state of Cr. The sketched bands (left side for spinup and right side for spin down) have many of the properties that were postulated for the intermediate band solar cell. Adapted from Ref. [107], Figure 8.3.

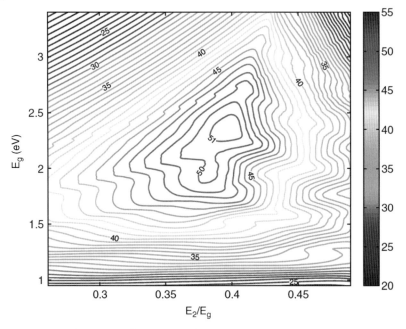

Figure 8.10 Photovoltaic efficiencies rise above 0.51 at one sun in theoretical investigation of the AlP:Cr system [107]. Adapted from Ref. [107], Figure 8.3.

structure diagram is split into two parts, the left for spinup and the right for spindown. The splitting of the narrow peak from the Cr ion indicates ferromagnetic order. The Cr plays the role of a "deep level" as was mentioned above in connection with doping of semiconductors. The Cr level is indicated as half-filled, so this band structure is similar to that needed for the intermediate band solar cell.

The authors point out that a single absorber material, rather than a multijunction system, helps avoid optical and tunnel junction losses, and generally might be simpler to construct than a competing tandem system. On the other hand, the particular compound has not yet been made, but its properties have been surveyed theoretically as a start. The parent semiconductor is AlP, in the zinc-blende structure with an indirect bandgap of 2.43 eV. The effect of random substitution of Cr on the Al sites was estimated theoretically. The material was found to be a dilute magnetic semiconductor of ferromagnetic type, with a critical temperature estimated at 160 K.

The efficiency of the resulting solar cell was found to exceed 50% in a suitable range of parameters, as shown in Figure 8.10. This is a very encouraging result for the proposed AlP:Cr cell, since the largest possible single bandgap cell efficiency is only 31%.

The optimum predicted behavior for photovoltaic conversion was 52% efficiency at one sun, at bandgap 2.43 eV, and E_2/E_G about 0.38. This remarkably high estimate is

Table 8.1 Photovoltaic conversion efficiencies [108].

Junction type	Cell efficiency	Module efficiency
Dye-sensitized solar cells	8.2%	3–5%
Amorphous Si (multijunction)	13.2%	6–8%
Cadmium telluride (CdTe) thin film	16.5%	8–12.8% (http://www.bizjournals.com/denver/news/2011/04/07/ge-to-buy-solar-panel-maker-primestar.html.)
Copper–indium–gallium–selenides (CIGS)	19.9%	9–15.8% (http://www.semiconductor-today.com/news_items/2011/SEPT/AVANCIS_070911.html.)
Polycrystalline silicon	20.3%	12–15%
Monocrystalline silicon	23.4%	14–16%
High-performance mono-crystalline silicon	24.7%	17–20%
Triple junction (GaInP/GaAs/Ge) 250 suns	40.7%	
Triple junction (GaInP/GaInAs/Ge) 454 suns	41.1%	

based on an advanced calculation. The authors point to a large literature related to dilute magnetic semiconductors, which might be transferable to solar cell applications.

8.4
Efficiencies: Three Generations of Cells

In discussing efficiency, it is useful to separate results on small laboratory junctions and those in panels, arrays of junctions (as available) of size measured in meters rather than centimeters. A summary of the most important solar cells and panel types [108] is given in Table 8.1.

From the table one sees that the new concepts, beyond the multijunction cells, have not reached the marketplace. The table is incomplete in that module efficiencies for the SolFocus arrays of concentrating cells are not included. It appears that the advanced organic tandem cells described in Section 7.1.3 have not entered the market, while the dye-sensitized cells have done so, and are available also as panels. The worldwide production of solar cells is reported 15.9 GW in 2010 (http://pvinsights.com/Report/ReportPMM04A.php.), with the top five producers (Suntech, First Solar, Sharp, Yingli, and Trina Solar) each exceeding 1 GW in production. Of these companies, First Solar is a leader in producing CdTe panels based on thin-film cells. The other firms, mostly located in China, sell polycrystalline or monocrystalline Si panels. As we will see later, production overall is rapidly rising, with most of these producers doubling their output in 2010 over the previous year.

9
Cells for Hydrogen Generation; Aspects of Hydrogen Storage

9.1
Intermittency of Renewable Energy

Hydrogen is available, but tightly bound, in water and many other compounds. To produce hydrogen from water requires an energy input. Hydrogen is not a source of energy, but rather a means of storing energy. A hydrogen infrastructure for storing and transporting energy is a possible alternative to batteries to store energy from solar cells. In a large-scale scenario, liquid hydrogen might be transported by railroad tank car, analogous to liquid natural gas (LNG), rather than by electric transmission over power lines, which are costly and require connected plots of land. One, but not the only, clear route to produce hydrogen from solar energy is via photovoltaic (PV) cells followed by electrolysis of water. We can take the point of view that the viability of this route is contingent on availability of large-scale inexpensive solar power, which the CIGS and CdTe cells seem to offer.

9.2
Electrolysis of Water

Inert electrodes such as stainless steel or platinum, immersed in water (containing a small addition of ions to promote conductivity), will evolve hydrogen gas at the cathode and oxygen gas at the anode if $\sim$1.9 volts is applied. The chemical potential energy associated with a molecule of hydrogen is 1.23 eV, so the efficiency of the electrolyzer is stated as 65%. According to Turner [109], the efficiency of commercial electrolyzers is in the range 60–73%. (Actually, about 4% of commercial hydrogen production is by electrolysis, and about half that is by electrolysis of brine (NaCl plus water) with chlorine gas, the primary desired product, hydrogen sometimes being abandoned.)

If the electrolyzer is connected with a photovoltaic array (PV) of efficiency 12%, connecting enough PV cells in series to achieve a working voltage greater than 1.9 V, then the overall conversion efficiency, sunlight energy into chemical energy in the form of hydrogen, is 0.65 × 0.12, which gives 7.8%. This is an off-the-shelf approach with commercial products available at present. A tandem cell is needed to reach the

Nanophysics of Solar and Renewable Energy, First Edition. Edward L. Wolf.
© 2012 Wiley-VCH Verlag GmbH & Co. KGaA. Published 2012 by Wiley-VCH Verlag GmbH & Co. KGaA.

minimum 1.9 V, and this, as noted in Chapter 7, section titled, "Low-Cost Tandem Technology: Amorphous Silicon:H-Based Solar Cells," is available in amorphous silicon tandem cells, providing 2.2 V.

We will see below an example of a monolithic PV-hydrogen electrolytic converter with efficiency 12.4%. For comparison, a high-temperature "solar thermal" water-splitting process is predicted to provide 24% efficiency [110] in hydrogen production but requires a large facility.

In general, we regard cells for electrolysis of water, fuel cells, batteries, Stirling engines, and electrical generators as off-the-shelf items, and will focus here on research topics that might eventually broaden the set of available devices.

9.3
Efficient Photocatalytic Dissociation of Water into Hydrogen and Oxygen

One element in this category is a cell for splitting water to produce hydrogen with sunlight, called photocatalytic splitting of water. Direct photocatalytic water splitting is not easy to accomplish. Efficiency is needed in the several steps: absorption of photons to create electron–hole pairs, in their transport to the water interface, in dissociating the water, and in separating and collecting hydrogen. The large area needed for a useful amount of hydrogen production, stemming from the low energy density in sunlight, is a difficulty, which could be addressed with focusing mirrors or Fresnel lenses. We consider a device in Figure 9.1 [111] based on a tandem solar cell.

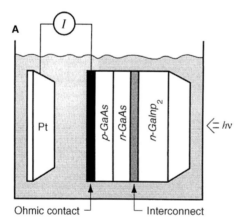

Figure 9.1 Schematic diagram [111] of illuminated monolithic photovoltaic–photoelectrochemical device for hydrogen production by water splitting, with 12.4% efficiency in converting solar light energy into chemical energy. In this cell, with no external connection, hydrogen bubbles appear on the illuminated surface at the right (cathode), and oxygen bubbles appear at the left (platinum anode). Mass spectroscopy revealed the evolved gases to be pure hydrogen and oxygen in ratio 2 to 1. In this device, electrons flow to the illuminated contact where they act to release hydrogen gas. This cell is built on the GaAs multijunction tandem solar cell technology as described in Chapter 7, and is clearly an expensive device.

9.3 Efficient Photocatalytic Dissociation of Water into Hydrogen and Oxygen

This device [111] can be regarded as a water electrolysis cell in series with a tandem solar cell, to supply the needed voltage to split water. Crudely, one can think of the semiconductor layers as a (light-activated) battery, the right-hand cathode terminal is in the electrolyte and the left-hand terminal ("ohmic contact") is insulated from the electrolyte, but connected to the platinum (foil or gauze) anode through the external ammeter. The electrolyte in the cell is 3 molar H_2SO_4, a strongly acidic solution conductive by H^+ ions, also known as solvated protons or "hydronium ions" H_3O^+.

Consider first the electrolytic cell aspects. The cathode was coated with a thin layer of platinum particles, in the nature of "platinum black," by an electroplating procedure.

The cathode reaction (reduction) is

$$2H^+ \text{ (aq.)} + 2e^- \rightarrow H_2 \text{ (gas)}$$

The anode reaction (oxidation) is

$$2H_2O \text{ (liq.)} \rightarrow O_2 \text{ (gas)} + 4H^+ \text{ (aq.)} + 4e^-$$

Taking twice the first reaction and adding to the second reaction, we get

$$2H_2O \text{ (liq.)} \rightarrow O_2 \text{ (gas)} + 2H_2 \text{ (gas)}$$

The standard potential for this reaction is 1.23 V. To make the reaction work, a larger voltage is needed, the excesses being termed cathodic and anodic overvoltages.

Solar cell output voltages are limited by the bandgaps of the underlying semiconductors, which are 1.83 eV on the right and 1.42 eV on the left, in Figure 9.1. The sum of these voltages provides an upper limit of voltage from the tandem cell, which well exceeds the voltage, nominally 1.23 V, to decompose water.

The photocurrent flows along the wire at the top of the diagram (ammeter was shown in the previous diagram) and the Fermi levels of the ohmic contact and the platinum electrode are aligned. So, the voltage developed in the semiconductor layers is dropped across the electrolye, driving the decomposition of water. In the two semiconductor layers, it is seen that the Fermi level is flat (no voltage drop) across the connecting "tunnel diode interconnect," so that the photovoltages of the two junctions are additive. The light enters on the right, and the right-hand junction has the larger bandgap, so that lower energy photons that are not absorbed on the right pass to the left to the second junction where they are absorbed.

In such a tandem cell, the electrical current has to be the same in each junction, so that the photocurrent of the series tandem cell is less than the short-circuit current of the weaker junction. The voltages add directly, allowing a higher conversion efficiency than a single-junction photocell. The efficiency was calculated as the chemical energy in the resulting hydrogen divided by the radiation power input. The chemical energy was calculated from the measured current multiplied by the voltage 1.23 V that is related to the hydrogen molecule.

The efficiency of water splitting was calculated from curve 1, with no external voltage applied, as power-out divided by power-in, or 0.12 A × 1.23 V/(1.190 W) = 0.124, for 1 cm^2 (Figure 9.3). The light intensity was about 11 suns for this experiment. (The

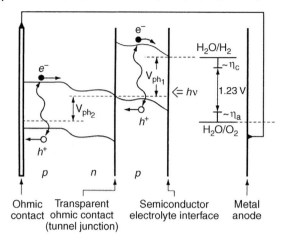

Figure 9.2 Schematic band diagram [111] (electron energy increasing vertically) of illuminated monolithic photovoltaic–photoelectrochemical device for hydrogen production by water splitting, with 12.4% efficiency. (This is the same device as shown in Figure 9.1, but the platinum anode has been moved to the right side of the picture.) In this device, electrons flow to the illuminated semiconductor–electrolyte interface, where they act to release hydrogen gas. Energy levels for water reduction (release of hydrogen) and water oxidation (release of oxygen) are sketched on the right side of diagram. An alignment is needed between conduction band edge and water reduction level, at the semiconductor interface. Alignment is also needed between the lower water oxidation energy and a level providing holes from the anode. So, on the right, the metal Fermi level (dashed line on the right) has to be pushed down by 1.23 eV + 2 η to allow a hole to be injected into the water (i.e., platinum anode accepts an electron from the water) and thus to release oxygen.

device here described is expensive and not a candidate for any practical water splitting approach. In the next sections, we will find examples of cheaper tandem cell possibilities.) One can infer that adding an additional junction in the cell to make a three or more junction tandem cell could result in excess cell output voltage beyond that needed to release hydrogen. Such a cell might be configured to simultaneously split water and supply power to an external load.

9.3.1
Tandem Cell as Water Splitter

The situation (Figure 9.3) in curve 2 (zero photocurrent at zero applied voltage, that is, the water splitter does not work) is usually the case for a single junction. The reason is that the conduction and valence band energies must be separated by 1.2 eV or so for efficient absorption of solar spectrum photons. At the same time, the conduction band edge has to be above the water reduction energy level and the valence band edge has to be below the water oxidation energy level (these two levels, separated by 1.23 eV + 2 η ~ 1.6 eV, are shown in Figure 9.2 on the right). It is common to reduce the overvoltage losses η at the semiconductor/electrolyte interfaces by depositing platinum particles as catalysts. (We will see below that recently progress has been

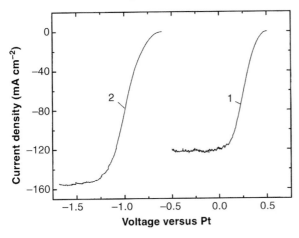

Figure 9.3 Current–voltage diagrams [111] of tandem cell (right, curve 1) and single right-hand cell (*p*-GaInP$_2$ cell) under white light illumination. A voltage source has been inserted in the upper connecting wire as shown in Figure 9.2, to obtain curve 1. The voltage source has been inserted between the platinum electrode and the inner ohmic junction (labeled "transparent ohmic contact") to obtain curve 2. Note that the zero of current is at the top of the figure. At zero voltage, a current of 120 mA flows through the tandem cell, but to get a current from the (illuminated) single cell (curve 2), a voltage has to be provided externally. The efficiency of the cell is calculated from curve 1 assuming that energy 1.23 eV is associated with each hydrogen molecule.

made in finding less expensive catalysts for this purpose.) The platinum deposition is usually done by electrolytic deposition from a solution containing platinum ions. These conditions cannot usually all be satisfied at once without inserting an external voltage, which has been done in the present example by the extra PN junction. A survey of the energy levels is given in the next figure.

According to the compilation of Figure 9.4 [112], GaAs, GaP, and TiO$_2$ are suitable to act as photocathodes to release hydrogen from water, in that their conduction band edges all lie above zero on the "NHE," normal hydrogen electrode, scale.

9.3.2
Possibility of a Mass Production Tandem Cell Water-Splitting Device

Shown in Figure 9.5 is a schematic of a potentially low-cost tandem device that achieves direct cleavage of water into hydrogen and oxygen by visible light. This is based on two photosystems in series (tandem), with electron flows as shown in the figure. The top cell (exposed to light) is a thin film of tungsten trioxide that absorbs the blue portion of the solar spectrum. The valence band holes decompose water directly to oxygen. The conduction band electrons are fed into the second photosystem, the dye-sensitized nanocrystalline TiO$_2$ cell. This is placed directly under the tungsten trioxide film and captures the green and red part of the solar spectrum. The photoelectrons in the conduction band of the titania reduce water to produce hydrogen gas.

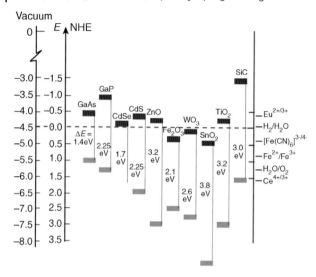

Figure 9.4 Survey [112] of energy gaps and band-edge energies of semiconductors and oxides of use in solar cells and water splitters. Note that the reduction energy level of water, taken as zero on hydrogen electrode energy scale of electrochemistry, is about 4.5 eV below the vacuum energy. On the right are shown energy levels for some additional reduction/oxidation reactions in water solution.

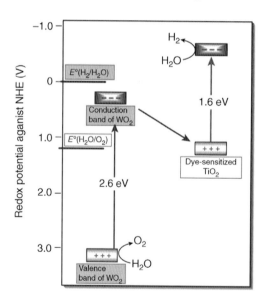

Figure 9.5 Proposal [112] for inexpensive thin-film tandem cell for water splitting. In its realization, the two cells are stacked, just as in Figure 9.1. The downward sloping arrow in the center of this figure represents the "transparent ohmic contact" connecting the two cells in Figure 9.1. The "dye-sensitized TiO_2" cell on the right is as described in connection with Figures 6.16–6.18, and the 1.6 eV arrow shown here corresponds to the dye absorption energy rather than the titania band gap.

This device perhaps can be produced as a wide-area membrane with water on both sides, generating oxygen on the one side and hydrogen on the other side. Unlike the expensive semiconductor system discussed earlier, the oxide surfaces are inert with respect to water exposure and do not involve single crystals or vacuum processing steps.

9.3.3
Possibilities for Dual-Purpose Thin-Film Tandem Cell Devices

One might imagine a monolithic dual-purpose self-regulating photoelectrochemical device comprised of a multijunction tandem solar cell, electrolytic cell, and power sharing circuitry. The tandem cell structure at illumination will be efficient inherently by using several different gaps. Its series structure will give an open-circuit voltage larger than that necessary to decompose water. The excess photovoltage with respect to water splitting can be diverted to an external load. The cell will generate primarily electric power under heavy load (small load resistance) and will generate primarily hydrogen gas at light load (high load resistance). The structure might be built up as a CIGS cell on aluminum foil, with, for example, a tungsten trioxide blue light cell deposited on top. The upper surface of the tungsten trioxide cell could be metallized with a transparent electrode, and fed through a load resistor to a second aluminum foil below the original aluminum foil, the interspace between the two foils being filled with water to be electrolyzed.

9.4
The "Artificial Leaf" of Nocera

The type of cell shown in Figure 9.1 has been updated recently by Daniel Nocera [113] (http://techtv.mit.edu/videos/633-daniel-nocera-describes-new-process-for-storing-solar-energy) in three important ways. First, he has demonstrated that a commercial tandem silicon solar cell, such as described in Figure 7.12, which can be made inexpensively in flexible thin-film form, can replace the expensive GaAs multi-junction cell. Second, he has demonstrated cobalt phosphate-based catalysts that replace the expensive platinum catalyst films described in Figure 9.1. Third, Nocera described the photocatalytic device as an "artificial leaf," pointing out its basic nature and connection with photosynthesis. Photosynthesis includes water splitting in the set of reactions that use CO_2, H_2O, and photons to produce sugar and carbohydrates.

To quote the author [113], "The MIT team spread its catalysts on opposite sides of a silicon wafer (tandem cell). The silicon absorbs sunlight and passes energetic, negatively charged electrons and positively charged electron vacancies to the catalysts on opposite sides that use them to make H_2 and O_2. When the device is placed in a clear jar and exposed to sunlight, the setup converts 5.5% of the energy in sunlight into hydrogen fuel." This almost allows water splitting on a small scale to be implemented using off-the-shelf components, since the silicon tandem cells have been made in large quantities for some time for application as rooftop tiles. In addition, Nocera has taken steps toward commercializing this for rooftop installation. The suggestion is that every house or small neighborhood in a village might have its own energy supply. It appears that a fuel cell would be used to turn the stored hydrogen into electricity when needed. A small-scale solution applied to billions of homes potentially becomes a large-scale solution. The suggested analogy may be the mobile cellular phone, available on an individual basis to billions of people. It is reported (http://en.wikipedia.org/wiki/Mobile_phone) that in 2010 the number of subscriptions to mobile phones was 4.6 billion. Nocera advocates "bottom-up" solutions of the projected energy shortage (http://techtv.mit.edu/videos/633-daniel-nocera-describes-new-process-for-storing-solar-energy) and is critical of the conventional power grid, which is prone to system-wide failure and does not reach locations where as many as 1.4 billion people live. Fuel cells, which turn hydrogen into electricity, are high technology and not inexpensive. Conventional methods for storing hydrogen range from high-pressure tanks to liquid tanks, but convenient small-scale methods are still in development, as we will discuss next.

The publicity on the "artificial leaf" seems to emphasize the small-scale, hand-held device, but to be useful it would seem that a collection area perhaps $10\,m^2$, about 10.3 feet on a side, might be minimum useful area for a roof. If electrical efficiency is 10%, on a day/night average sun power of $200\,W/m^2$, this would average 200 watts on a continuing basis, while the peak power would be $1000\,W/m^2 \times 0.1 \times 10\,m^2 = 1000\,W$. According to the estimate above of 5.5% conversion of sunlight energy $200\,W/m^2$ into its hydrogen equivalent, this would give an average power about 110 watts in hydrogen equivalent. To apply the rule-of-thumb cost of \$1/W peak power, at peak power 1000 W the cost estimate is \$1000 for the rooftop array. That is a minimum for the cost of the solar cells alone, ignoring the water circulation, hydrogen collection, and fuel cell aspects. If the system were allowed to accumulate hydrogen over a day, the hydrogen energy would be 110 watts $\times 24 \times 3600\,J = 9.5\,MJ$, or about 3.6 kWh. The nominal value of the energy at 14c/kWh is then \$0.5 per day, or \$184 per year. On this basis, it would take 5.4 years to pay back the assumed \$1000 investment.

While the emphasis of Nocera [113] (http://techtv.mit.edu/videos/633-daniel-nocera-describes-new-process-for-storing-solar-energy) is on small-scale applications, where the storage of the resulting hydrogen seems problematic, on a large-scale photovoltaic installation "artificial leaf" cells might also be useful. Assuming the cells can automatically switch from delivering power to an electrical load to producing and collecting hydrogen based on the load demand, a large-scale facility could afford even a liquefying system to store hydrogen produced when the electrical demand is small.

9.5
Hydrogen Fuel Cell Status

A fuel cell takes hydrogen gas as input and produces electricity, with efficiency in the range of 50–70%, greater than the internal combustion engine, ~30–35%.

Automaker Honda in 2008 [114] released its hydrogen-fuel cell powered sedan, the FCX. This vehicle has a 280 mile range, using pressurized hydrogen. Pressurized hydrogen is available only in a few places, for example, at $5/kg at a series of stations in California. This auto is expensive, largely owing to the cost of the fuel cell. It accelerates rapidly and can reach 100 mph. The manufacturer is said to have a production line available, which it can run when the infrastructure of hydrogen stations expands. At the moment, the manufacturer is leasing these cars at a low introductory rate. This car has a fuel cell "the size of a desktop PC that weighs about 150 pounds" and has a power rating of 100 kW (134 hp, at 746 W/hp). It is anticipated that the cost of the fuel cell unit will fall, so that the cost of producing the car will fall from "several hundred thousand dollars" to below $100 000. The manufacturer says that the car is efficient, "equivalent of 74 miles a gallon of gas." Other manufacturers sell cars that simply burn hydrogen gas in an internal combustion engine, an expedient that loses the efficiency inherent to the fuel cell/electric motor combination, and ignores the release of carbon dioxide.

9.6
Storage and Transport of Hydrogen as a Potential Fuel

Gaseous hydrogen can be piped as is natural gas, with preference for plastic pipe to avoid questions of hydrogen embrittlement of steel. Large volume storage of hydrogen (as is presently the case for natural gas) may be possible in favorable geological underground formations. Caverns presently sought for storing or "sequestering" unwanted CO_2 underground may conceivably also prove [115] useful alternatively for storage of hydrogen gas or as large pressure vessels to store energy, if fitted with compressors and exhaust turbines to generate electricity.

Liquid hydrogen requires the low temperature of 20.3 K. Liquid hydrogen has high energy density per unit mass, but on a volume basis the energy density is about a factor 4 less than gasoline or diesel fuel. Storage tanks for liquid hydrogen, which are vacuum vessels with "superinsulation" in the form of 30–300 layers of aluminized mylar, are available in a variety of sizes, including 100 l and 190 l. These tanks accommodate hydrogen liquid at atmospheric pressure, without internal refrigeration and have losses in the range of 1% per day, decreasing with increasing tank volume. A line of sedans has been developed and tested by manufacturer BMW that use such liquid storage tanks, the gas powering internal combustion engines.

The schematic in Figure 9.6 [116] of a proposed process cools hydrogen at atmospheric pressure in a series of three heat exchangers, each heat exchanger block being cooled by expansion of a mixture of neon and helium gas. The proposed cycle would require only one pass of hydrogen through the system, while existing cycles require multiple passes to liquefy all the gas. The energy and capital expenses in liquefaction are not negligible. Commercial liquefaction schemes usually involve compression of the hydrogen gas.

Tank trucks are on the highway transporting liquid helium, and other high-value cryogenic liquids including liquid oxygen and liquid nitrogen. Liquid natural gas is

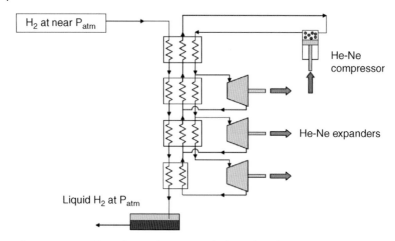

Figure 9.6 Possible single-pass low-pressure hydrogen liquefaction cycle, proposed [116] by Gas Equipment Engineering Corp. In this diagram, the resistor-like symbols are heat exchangers, and the cooling occurs as the He–Ne gas mixture, initially pressurized, is allowed to expand.

transported on a larger scale by ship and railroad tank car. On a large scale, hydrogen transport as a liquid is economic for some applications.

On the smaller scale, to replace the gas tank of an automobile, liquid storage may be a less viable option, since the fuel would disappear over relatively short storage times. At present, the most common hydrogen storage is in a high-pressure tank.

9.7
Surface Adsorption for Storing Hydrogen in High Density

Alternative possibilities for dense hydrogen storage are by reaction of hydrogen to form chemicals such as metal hydrides and by temperature-dependent adsorption of hydrogen onto lightweight high-area substrates, which may include decorated graphitic surfaces. Hydride storage tanks have been tested in connection with autos and trucks. A figure of merit of such a storage medium is the weight density of hydrogen that can be achieved. A practical benchmark of 6.5% by weight of hydrogen has been chosen by the U.S. Department of Energy.

A large surface area on which hydrogen adsorbs with a small binding energy is suggested in the cover image (lower left) of this book. The concept shown is a nanoporous structure made purely of carbon, using graphene planes and carbon nanotube pillars. Such a structure is not manufacturable at present, but illustrates properties that are desired. The weak adsorption energy allows hydrogen to be desorbed by a small increase in the temperature, and the surface density is clearly high in such an imagined structure.

A simpler storage model based on graphene layers is shown in Figure 9.7. The modeling associated with Figure 9.7 suggests that graphene layers optimally spaced at 0.8 nm may, with some uncertainty, accommodate hydrogen at the 6.5 wt % level

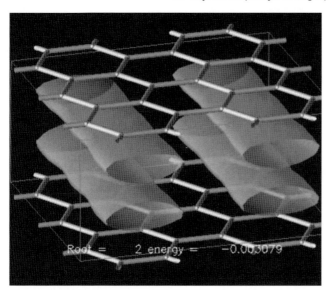

Figure 9.7 Nanometer-scale schematic diagram [117] of one hydrogen storage configuration examined theoretically. The tubular regions, located between graphite planes of variable spacing, are regions of high probability density for hydrogen molecules. The modeling is based on van der Waals attraction.

at room temperature and moderate pressure. This structure could be described as graphite intercalated with hydrogen. The calculations [117] indicate the hydrogen molecules will be delocalized (i.e., able to freely move, between the graphitic layers), which would promote rapid filling and emptying of the charge. On the other hand, the spacing of the graphene layers is a critical parameter that has to be arranged independently. No obvious route to obtain the desired spacing of graphene layers is known, although there is a large literature of "graphite intercalation compounds." Any intercalation compound will have additional mass and will block free volume for the intended hydrogen molecules.

The attraction of hydrogen to transition metal atoms, including Ti, is stronger than the attraction to graphene as discussed above. According to Durgun *et al.* [118a], the bonding strength is about 0.4 eV, which is compatible with adsorption/desorption at room temperature. The interaction can be reliably calculated with advanced methods, leading to predictions, indicated in Figure 9.8, of hydrogen storage in certain simple organic molecules with transition metal adsorbates, in the range of 14 wt %.

Durgun *et al.* [118a] find that the hydrogen molecules in the right panel of Figure 9.8 are bound with energies varying from 0.29 to 0.49 eV, which are considered suitable for room-temperature storage of hydrogen. This is a reliable theoretical estimate, but synthetic chemistry is needed in order to create the underlying molecule, $C_2H_4Ti_2$. The molecules would need to be dispersed, in a state more like a gas than a liquid, if hydrogen is to be readily inserted and extracted upon demand. This leaves the problem that the molecules will not reliably remain in the storage

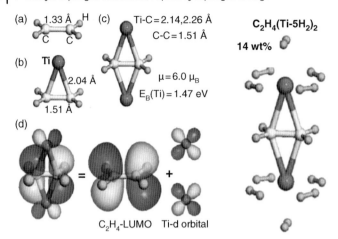

Figure 9.8 Diagrams [118] showing hydrogen storage molecules. (a) Structures based on ethylene C_2H_4; (b) indicates a single Ti atom with bond length 2.04 Å from carbon, followed by adsorption of two Ti atoms at binding energy 1.47 eV; (c) shows how this structure can be regarded as a covalent bond between the lowest unoccupied molecular orbital of the ethylene molecule (large lobes in (d)) with the 3D orbital of Ti (small lobes in (d)). Right-hand side of the figure shows binding sites of 10 hydrogen molecules that are provided by the two titanium atoms in this structure. This corresponds to 14 wt % of hydrogen.

tank, but might flow out as the hydrogen fuel is withdrawn. The authors suggest that the molecules should be embedded in a nanoporous matrix. The expected means to bring hydrogen gas in or out of the storage sites is to adjust the temperature. A nanoporous matrix might facilitate this, if it were continuous and could be uniformly heated by passing current through it.

It has been recently shown [118b] that the adsorption sites for H_2 shown in Figure 9.8 are blocked by even small amounts of O_2. These authors say that to make use of such designs, in practice, where some oxygen will always be present, access to the surface by oxygen would have to be blocked by a polymeric coating or by a nanoporous structure that would not allow molecular oxygen to enter. Palladium–silver alloys are permeable to hydrogen but not other gases, and could, in principle, block entry of oxygen to a cask containing the activated surfaces as shown in Figure 9.8. The idea of a porous structure that will allow a particular chemical element to enter and leave is suggested in Figure 10.9. This figure applies to Li ions, but a similar picture might apply to H_2.

Similar results for hydrogen storage in titanium-decorated polymers, notably polyacetylene $(C_2H_2)_n$, have been presented by Lee et al. [119]. (Whether oxygen would block these sites is not clear.) Polyacetylene forms sheets, so that containing the material in the "gas tank" while hydrogen is drawn in and out is easier. The authors say that titanium-decorated polyacetylene will store 9% by weight of hydrogen if hydrogen is admitted at 25 °C and initial pressure 30 atm, and is desorbed at 2 atm and 130 °C. These are practical conditions. The bonding geometries are similar to those shown in Figure 9.8.

9.7.1
Titanium-Decorated Carbon Nanotube Cloth

A solution seems to exist that builds upon aspects of the two previous figures. Titanium atoms directly bridging two carbon atoms can decorate graphene planes, either as sheets or as nanotubes. Durgun *et al.* [118a] and also Lee *et al.* [119] have previously predicted rather similar hydrogen storage results based on titanium atoms bridging carbon sites in carbon nanotubes. Carbon nanotubes can be formed as a sheet or "cloth" that is strong, lightweight, and electrically conductive [120].

It is assumed that such a cloth can be decorated with a high density of Ti atoms, bridging carbon sites as in the cases above, to serve as a nanoporous storage medium for hydrogen. Ti deposition could be accomplished on the cloth in a CVD reactor. Hydrogen storage with desorption by electrical heating could be based on such a carbon nanotube sheet, as indicated in Figure 9.9. This figure shows a carbon nanotube cloth operating as a planar incandescent lamp, to illustrate the ability to control its temperature by simple passage of current through the cloth.

Figure 9.9 Uniform incandescent light radiation demonstrates uniform electrical Joule heating of carbon nanotube cloth, dimensions 16 mm × 12 mm [120]. This sheet of multiwall carbon nanotubes is free-standing, attached to vertical supports that serve as electrical contacts. The sheet temperature was adjustable over the range 1000–1600 K. The low mass of the nanotube sheet resulted in a time constant for temperature change on the order of 0.1 ms.

These adsorption-based hydrogen storage methods, if the oxygen poisoning aspect [118b] can be overcome, seem adaptable to small-scale applications, such as hydrogen that might be produced by a rooftop system along the lines envisioned by Nocera.

9.8
Economics of Hydrogen

One scenario for future distribution of energy is in the form of hydrogen, which might be compressed or liquefied. The "hydrogen economy" scenario has been promoted along with the idea that automobiles may switch to electric motors for which the electricity comes from fuel cells run on hydrogen gas stored in the car's fuel tank. A scenario of this sort is being tested in Iceland, which has renewable energy resources and no oil, coal, or gas.

Hydrogen as a fuel is attractive because it burns to produce water, with no adverse effect on the atmosphere. Hydrogen is not abundant in its free form; for example, the availability of hydrogen as a byproduct of commercial liquefaction of air to produce nitrogen and oxygen is limited. Commercial (nonrenewable carbon-based) production of hydrogen is extensive. It is typically based on methane (natural gas), and approximately half is used to make ammonia for agriculture, the other main use being in oil refining. Worldwide production of ammonia [121] was 109 million metric tons in 2004, where a metric ton or tonne is 1000 kg. The final reaction is direct stoichiometric combination of hydrogen and nitrogen, to form NH_3 in a catalytic process called the "ammonia synthesis loop" (Haber–Bosch process).

9.8.1
Further Aspects of Storage and Transport of Hydrogen

Commercial storage and transport of hydrogen is conventionally at high pressure in steel cylinders, or aluminum-lined containers reinforced with carbon fibers, or as liquid in a cryogenic container. The network of hydrogen "filling stations" in California dispenses hydrogen at 5000 psi and at 10 000 psi (~330 bar to 660 bar). There are significant energy costs in pressurizing hydrogen and also in liquefying it. (In principle, hydrogen can also be stored at high density adsorbed to high surface area substrates, such as graphitic planes decorated with hydrogen adsorbing centers, such as Ti atoms, as mentioned above.) Finally, ammonia, NH_3, is easily liquefied, and then contains hydrogen at high density. Liquid ammonia, in spite of its toxicity, is routinely piped over large distances in connection with its use as a fertilizer. As mentioned above, the final step in ammonia formation from hydrogen and nitrogen is the Haber–Bosch process, a well-known commercial catalytic process carried out at pressures on the order of 100 atm.

It is reported that Iceland produces 2000 tons of hydrogen per year by electrolysis, using spare electrical capacity. Most of it is used indeed in ammonia production, some in running fuel cell buses and boats. Eventually, the price of natural gas may rise enough that a renewable route to ammonia production will be competitive. (The energy cost of the Haber–Bosch process is high, since it involves four passes through a bed of catalysts, at about 400 °C, each time releasing pressure from 100 atm and then repressurizing.)

Hydrogen for ammonia production comes from steam reforming of methane (natural gas). Therefore, (nonrenewable) hydrogen production for ammonia alone, worldwide, can be obtained from the ammonia output multiplied by the fraction of the mass of anhydrous ammonia represented by hydrogen. This fraction is $(3 \times 1.008)/[(3 \times 1.008) + 14.0067] = 0.1776$, where the mass of nitrogen is 14.0067 in atomic mass units. Therefore, the worldwide production of hydrogen (specifically the part associated with production of ammonia) in 2004 was 109 million metric tons $\times\ 0.1776 = 19.35$ million metric tons. (The proportion of this produced in the United States was 8.17%, or 1.58 million metric tons of hydrogen.) Commercial hydrogen production is also done from "town gas," methane plus carbon monoxide, which is a result of gasification of coal. All of these are carbon-dependent (nonrenewable) processes that release greenhouse gas. An independent estimate is world production of hydrogen (2004) is 50 million tons, about half for ammonia and about half for petroleum industry; "hydro-cracking" to produce light hydrocarbons from heavy hydrocarbons.

9.8.2
Hydrogen as Potential Intermediate in U.S. Electricity Distribution

As an upper limit on hydrogen that might be needed, we can calculate the hydrogen production equivalent in energy to the U.S. electricity usage per year, which was 460 GW in 2004 [122]. The total electric energy in 2004, then, taking a year as 3.154×10^7 s, was 1.45×10^{19} J. If the energy carried per H_2 is 1.23 eV $= 1.968 \times 10^{-19}$ J, and the mass per H_2 is $2 \times 1.008 \times 1.6605 \times 10^{-27}$ kg, then the equivalent mass of hydrogen per year is 2.47×10^{11} kg $= 2.47 \times 10^8$ metric tons, or 247 million metric tons. (Another estimate [109] is that the total transportation energy need of the United States could be met (2004) by 150 million tons of hydrogen.)

This estimate of 247 million metric tons of hydrogen is 12.8 times the worldwide production of hydrogen in the preparation of ammonia for fertilizer. The price per metric ton of ammonia on May 15, 2008 was quoted as $550, or $0.55/kg. If the cost of ammonia is attributed entirely to its hydrogen, which is 17.8% of its weight, then the cost of hydrogen is $0.55/(0.178) = $3.09/kg. It is clear that this is an overestimate because the costs associated with fractionating air to get nitrogen, and with the Haber reaction to make ammonia, would be avoided in the pure hydrogen production. An independent estimate of cost of hydrogen from natural gas is $2.70/kg. It is also stated that retail prices per kg for hydrogen at filling stations in California and in the Washington DC area range from $5 to $6 per kg. Again, according to Figure 1 in Turner [109], electrolysis of water would match $2.70/kg for producing hydrogen at electricity costs in the vicinity of 4 c/kWh. It is also stated that the lowest price hydrogen is produced by electrolysis at wind farms.

As the price of natural gas rises, it might occur that ammonia (fertilizer) production would represent a market for renewably produced hydrogen, for example, from electrolysis using electricity from wind farms or PV farms, Nocera photocatalytic artificial leaf devices, or thermochemical splitting of water. Water and nitrogen are regarded as renewable resources, effectively infinite in supply, while carbon is not renewable. (Turner [109] also states that potential electricity produced by solar and wind could, if the capacity were built, be more than adequate to produce all needs for hydrogen, based on electrolysis of water, and also that there is enough water for this purpose.)

Commercially available also, as we have mentioned, are fuel cells, with efficiency in the range of 50–70%, which produce electrical current from hydrogen gas. So a remote location might obtain continuous electrical power by the combination of PV, electrolyzer, hydrogen storage, and fuel cell. This is a viable approach, an alternative to the simpler combination of PV and batteries.

10
Large-Scale Fabrication, Learning Curves, and Economics Including Storage

10.1
Fabrication Methods Vary but Exhibit Similar Learning Curves

We are interested in the techniques needed to greatly increase the production and use of renewable energy. In terms of fabrication, it is well known, and makes common sense, that the cost of making a given device falls as methods of fabrication are improved with increased output. For photovoltaic modules, the cost per watt Wp over 30 years is summarized in Figure 10.1, plotted versus accumulated production (volume) in megawatts. The value Wp is the power the panel will produce under full illumination, 1000 W/m^2. The average cost is shown, including various cell types, but this curve primarily reflects silicon solar cells, which still represent 90% of production.

This nearly linear log–log plot of cost C per peak watt, Wp, versus manufactured volume V, is consistent with

$$C/C_0 = (V/V_0)^{-L}, \tag{10.1}$$

so that

$$\ln(C/C_0) = -L\ln(V/V_0). \tag{10.2}$$

The learning parameter L in this case is about 0.33. Note that one decade of cost, from \$10 down to \$1, occurs in about three decades of volume, from 100 MW to about 10^5 MW = 100 GW (which is extrapolated). As we saw at the end of Chapter 8, the volume in 2010 is reported as 15.9 GW (http://pvinsights.com/Report/ReportPMM04A.php), so the extrapolation seems reasonable.

On the Internet today, one can find many vendors (http://Wpww.wholesalesolar.com/solar-panels.html) of solar cell modules, with offered prices generally in the range \$3/Wp to \$5/Wp.

The price of silicon solar cells has recently fallen, reflecting increased production capacity of silicon and of silicon cells. A consequence is that smaller manufacturers of innovative products have faced losses. The September 2011 bankruptcy of the CIGS producer Solyndra [124] has been described as a cash flow problem arising from

Nanophysics of Solar and Renewable Energy, First Edition. Edward L. Wolf.
© 2012 Wiley-VCH Verlag GmbH & Co. KGaA. Published 2012 by Wiley-VCH Verlag GmbH & Co. KGaA.

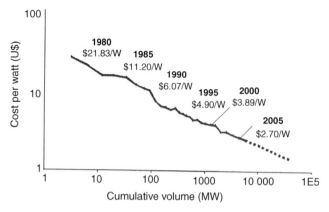

Figure 10.1 Photovoltaic module manufacturing learning curve [123]. Cost per watt plotted versus *cumulative volume* or *manufactured volume* of device capacity installed, in megawatts. The limit of $1/Wp corresponds approximately to 6 c/kWh electricity cost.

Solyndra's cost per panel of $3.94 versus the market selling price of $3.24. So it appears that today's price per panel is definitely below $4.00.

In a recent speech, Steven Chu, U.S. Secretary of Energy, spoke of reducing the cost of photovoltaic energy by 75%, to reach $1/Wp, which is regarded as competitive with other energy sources, and which in his speech was identified with electricity cost of 6 c/kWh [125]. (See Figure 5.6 for cost of various cell types. The cell types have been described in Chapter 6.)

On the other hand, the implied $4/Wp at present is higher than prices quoted on the Internet. $4/Wp in particular is much higher than $0.76/Wp, recently reported as the manufacturing cost for CdTe modules in large volumes [126]. The firm, First Solar, has stated their module efficiency as 11.2%, with reported production capacity of 59 MW per line. First Solar total production capacity is predicted to reach 2.2 GW in 2012, suggesting about 37 production lines. It was earlier reported that First Solar has an original line in the United States, 1 in Germany, and 16 in Malaysia.

A detailed history of the CdTe solar cell is available (http://en.wikipedia.org/Wpiki/Cadmium_telluride_photovoltaics). One of the advantages is the relatively simple *closed-space sublimation* process for depositing CdTe, which requires relatively low temperature and provides rapid deposition. Sublimation allows removal of molecules from a solid, without having to form a liquid. It appears that halogen lamps are sufficient to heat the bulk CdTe source, while it is mentioned that 500 °C is needed to process CIGS solar cells. So this makes clear that there are different learning curves for different technologies, and at present the CdTe technology is lower in cost than the predominant solar cell production, in silicon cells of several sorts.

See also "The Quest" by Daniel Yergin (Penguin, New York, 2011) for an anecdotal, business-oriented history of solar cells, Chapter 29, pp. 563–587, titled "Alchemy of Shining Light." Yergin is known as an explicator and apologist for the oil industry, see, for example, Daniel Yergin, "There will be oil," *Wall Street Journal*, Sept. 17, 2011.

This article suggests that of the total estimated oil in the earth in 1900, the fraction remaining is 0.58, or 1.4×10^{12} barrels. It is agreed that no new oil is being generated in the earth's crust, that is, that oil is not renewable.

We have mentioned solar cells of various types, starting with single-crystal silicon (Figure 6.5), multijunction cells using epitaxial layers of III–V semiconductors based on GaAs (Figure 6.7), polycrystalline and thin films of amorphous silicon (Figure 7.12), CdTe (Figure 6.15), copper indium gallium selenide (CIGS; Figures 6.11–6.14), dye-sensitized titania (Figures 6.16–6.18), and organic semiconductors (Figures 6.20 and 7.9). The concentrating cells (Figure 7.8) have to be provided, in addition, with mirrors or lenses, tracking mechanisms, and cell cooling. In general, each type of cell has a different set of fabrication methods, and it is likely that each method will have its own learning curve as its total production accumulates. One might imagine a more detailed version of Figure 10.1 with separate trajectories for separate technologies. The trajectory for CdTe clearly will be below the average curve shown.

To rank the fabrication methods in order of decreasing cost, it is likely that the liquid-phase epitaxy needed for the GaAs multijunction tandem cells is the most expensive. The crystalline Si solar cell is built on a wafer of Si, which is typically 200–300 μ thick, cut from a large crystal using a wire saw, which wastes also one wire-width of the Si single-crystal boule per wafer. A recent initiative has been reported to lower the cost of silicon cells by casting the wafers from molten silicon. This potential method avoids the crystal growing and sawing steps, saving both labor and raw material, and is projected [127] to reduce the cost by 50%. Cells made in this way are not yet on the market. It is not clear what quality and quality control can be achieved in such a process.

Polycrystalline silicon is conventionally deposited using chemical vapor deposition, and amorphous silicon using glow discharge. Large-size Si polycrystals, sometimes called multicrystals, can be quite efficient, because of the large grain size, which reduces undesired recombination (see Figure 6.8 and caption of Figure 6.4).

10.2
Learning Strategies for Module Cost

The cost per watt of product can be reduced by raising the efficiency of the cell. If one doubles the efficiency, then one needs only half as many modules to deliver the same power. Such savings apply both to the ingredients of the cells and, importantly, to the glass plates, frames, wiring between cells and between modules, and to the land on which the installation is placed. Improvements to cell efficiency can be analyzed from the physics of the cell, as discussed in Chapter 3, see Equations 3.69 and 3.70, and in Chapter 6, see Equations 6.5 and 6.6.

To summarize factors affecting cell efficiency, the power output is roughly the product of the open-circuit voltage

$$V_{oc} = (k_B T/e)\ln(J_L/J_{rev} + 1) \tag{3.69}$$

multiplied by the short-circuit current density, J_L. This $J(V)$ curve and its change under illumination are illustrated in Figure 3.17. Design and processing of the cells thus need to optimize the short-circuit current J_L and to reduce the reverse current, J_{rev}, which is seen from Equation 3.69 to enhance the open-circuit voltage, V_{oc}. Reducing the reverse current is aided by reducing the operating temperature (see Equation 3.67).

Learning curve events related to lowering the temperature are seen in Figure 7.8 in which an optical rod is used to block heat from the cell, letting through the active spectrum, and in Figure 7.17, where water-cooling is easily arranged when the concentrating cells are mounted on pontoons floating on water. Optimizing J_L means letting all the light into the cell, by using antireflection coating and trapping photons in the cell by surface texturing (see Figures 6.4 and 6.5), making sure the cell's absorber layer is thick compared to the absorption length (see Figure 6.3), and, importantly, by reducing recombination so that the carriers can flow around the external circuit. Reducing recombination means keeping the photogenerated minority carriers away from surfaces such as grain boundaries and the outer surfaces of the cell, and by passivating such surfaces to inhibit recombination, typically by reacting the cell with hydrogen gas, especially to fill dangling bonds in silicon cells. Other schemes for reducing recombination are shown in Figures 6.5 and 6.7a: oxidizing the surface facing the metal contact electrode, and arranging a back surface field to reflect minority carriers from the surface. In the development of the CdTe cell [126] (http://en.wikipedia.org/Wpiki/Cadmium_telluride_photovoltaics) (Figure 6.15), learning steps included reducing the thickness of the CdS layer to let more light through into the CdTe layer, and recrystallizing the CdTe layer by using $CdCl_2$, to increase grain size, which reduces recombination.

While each cell process has its own prescriptions, in general lowering thickness of layers reduces material use and increases the flow rate along the production line. It is clear that choice of process to minimize capital cost, such as avoiding evaporation and sputtering for deposition, which require vacuum equipment, is desirable, but may lead to trade-off with efficiency. A clear strategy in the support structure is to reduce the use of steel, replacing it by cheaper plastic supporting members such as PVC pipe in Figure 7.17.

It has been suggested that the nanoink process should benefit from a relatively simple and inexpensive deposition, but it seems that this process has not been very successful, as we will see in Figure 10.2. Lack of success can have different causes, including poor management of a company, but it also may be that formulating large volumes of the ink nanoparticles, whose composition must be accurately specified, may be difficult and expensive to accomplish and to control (http://Wpww.greentechmedia.com/articles/read/Thin-FIlm-Solar-Startup-Solexant-and-its-New-CEO/).

A part of the learning curve for CdTe producer First Solar, not mentioned in their press release, may be moving the bulk of their production to Malaysia, where labor is cheaper and where rules, for example, on disposal of waste, may be less onerous. These production lines are not intensely labor-intensive, however, which is a rueful point for the U.S. government, for which generating employment has been a goal.

Top 15 Thin-Film PV Producers, 2010

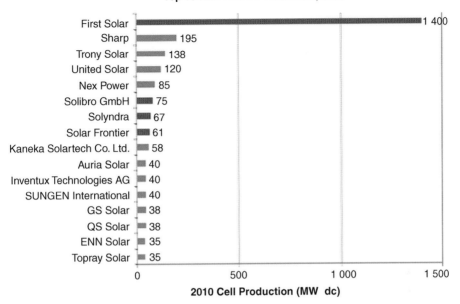

Figure 10.2 Thin-film solar cell manufacturers (http://Wpww.greentechmedia.com/articles/read/Thin-FIlm-Solar-Startup-Solexant-and-its-New-CEO/) in rank of production, in MW, in 2010. First Solar, the leader in thin-film solar cell production, makes CdTe cells. Sharp produces amorphous silicon cells, as do United Solar and Kaneka. Solar Frontier and Solyndra (bankrupt in Sept. 2011) and Avancis (not shown) produce CIGS cells. CdTe cells made by First Solar completely dominate the thin-film solar cell market at present.

Even so, it is reported [124] that the U.S. solar industry employs more people than either the coal or steel industries.

Another general learning approach is to use a wider foil or glass along the production line, because the expense of manufacturing equipment increases less than linearly (http://en.wikipedia.org/Wpiki/Cadmium_telluride_photovoltaics) as the width increases!

10.3
Thin-Film Cells, Nanoinks for Printing Solar Cells

The thin-film cells rationally seem to offer the best chance for matching the market prices in the large electric power market when the rising price of fossil fuels, or their adverse effects, curtail the conventional coal- and gas-fired power plants. A summary of the production of thin-film solar cells in 2010 is shown in Figure 10.2.

A range of fabrication methods are represented here. It appears that First Solar, which was reported to make CdTe modules at a cost of $0.76/Wp, uses a closed space sublimation method, starting from bulk CdTe. It is reported that their production now includes 16 fabrication lines in Malaysia, expanding from initial facilities in the United States. It appears that this is a simple and rapid fabrication of the basic CdTe absorber layer. It is reported (http://Wpww.greentechmedia.com/articles/read/Thin-FIlm-Solar-Startup-Solexant-and-its-New-CEO/) that Nanosolar and ISET (not in the first 15 producers) make CIGS cells by a nano-ink process, and that Solextant is starting to make CdTe cells using a nano-ink process. Other firms presently are using electrodeposition and reactive sputtering. Other CIGS producers are Avancis, Solar Frontier, Global Solar, and Honda Soltec; other CdTe producers are Antec, Abound Solar, and Primestar, now a part of General Electric. A variety of technologies are represented here, not all proven. For example, Solyndra, listed above as producing 67 MW of CIGS cells in 2010, went bankrupt in September 2011 [124]. Their innovation, a cylindrical geometry for the cell, to catch a range of sunlight angles, was too expensive to produce, as might have been clear from the start.

All of these processes involve production lines. A roll-to-roll process is used by United Solar for flexible multijunction Si cells, and by Nanosolar for its nano-ink process. An illustration of a roll-to-roll production line for Si is shown in Figure 10.3.

In other cases, the production line carries glass panels, which are coated sequentially. Figure 10.4 shows the production line of Abound Solar for CdTe panels (http://Wpww.abound.com/solar-modules/manufacturing), said to be 12.8% efficient.

The facility shown in Figure 10.4 accepts one glass panel every 10 s, which emerges at the other end of the line 2 h later as a completed CdTe solar module (http://Wpww.abound.com/solar-modules/manufacturing). At this rate, a 10 h day produces $8 \times 3600/10 = 2880$ modules, for a nominal daily volume of 0.288 MWp (at nominal 100 watts per panel). On this basis, 226 production days are needed to reach 65 MW, the stated annual capacity of the production line. This firm has funding for expansion of manufacturing capacity by 180 MWp in Colorado and an additional 600 MWp in Indiana. Exemplary panels are said to have 12.8% efficiency. This is the kind of company for which the future will depend upon the manufacturing cost, which is not easily available.

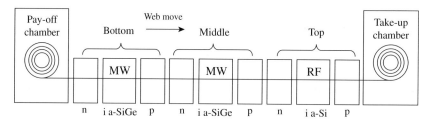

Figure 10.3 Roll-to-roll processing line using CVD deposition illustrated for three-layer silicon cells [128]. The CVD processing stations here include microwave and radio frequency heating. This is similar to a line used by United Solar (see Figure 7.12), where the moving foil or web is stainless steel, 35.6 cm in width, as mentioned following Figure 7.13. A simpler roll-to-roll scheme is used by Nanosolar, with ink deposition for the CIGS absorber layer. In both cases, the foil is later cut up and mounted in modules, extra steps that add to the cost.

Figure 10.4 Production line for CdTe panels in Longmont CO, USA, by firm Abound Solar (http://Wpww.abound.com/solar-modules/manufacturing). This line is characterized as capable of 65 MWp annual capacity.

10.4
Large-Scale Scenario Based on Thin-Film CdTe or CIGS Cells

Our simple analysis to address the full electric power consumption in the United States includes only solar cells, while in a realistic change to renewable energy there would likely be important contributions from wind turbines and from solar thermal plants, also called CSP (concentrated solar power).

(CSP is rapidly being built in the United States. Under construction are the Ivanpah, CA, power tower, 370 MW, Solana, Arizona, parabolic trough 280 MW and 6 h molten salt heat storage, and Crescent Dunes, Nevada, power tower, 110 MW with 10 h storage (http://en.wikipedia.org/wiki/List_of_solar_thermal_power_stations). Announced CSP projects in United States total 7 GW. These plants are either similar to the facility shown in Figure 5.4 or based on parabolic trough reflectors, and require cooling water for the steam turbine generator. It appears that a 200–280 MW plant will require 1500–3000 acre-feet of water per year. As we discussed in Chapter 5, the conversion efficiency of these plants is generally higher than photovoltaics, with the possibility of higher efficiency if turbine temperatures can be increased.)

Our present estimate is based on photovoltaic cell arrays, with no proviso for energy storage or need for water-cooling on the site. To provide 460 GW on average, since the sun shines only a fraction of the day, the capacity has to be multiplied by about $1000/205 = 4.9$, with the implication of a huge amount of grid storage, drawn upon to provide power at night. Our present analysis is not realistic, but provides an idea of the magnitudes involved in the PV part of what more realistically will become a combination of PV and wind power sources.

To supply full U.S. electric power from solar cells, the most reasonable approach would seem to be via CdTe or CIGS thin-film cells that have a cost advantage both in principle and in practice. These materials both have high absorption coefficients, meaning the absorber layer can be thin on the order of 3 μm, and both can be fabricated in economical production lines. The leading thin-film approaches, in our opinion, are CdTe cells (such as made by First Solar and Abound Solar), and CIGS $CuIn_{1-x}Ga_xSe_2$ cells (in spite of the present low market share of CIGS cells), such as produced by Honda Soltec in Japan, Nanosolar in California, and Avancis (in Germany and South Korea). The cost per watt, Wp, is certainly less than $1 in each case [126, 129].

U.S. electric power usage in 2004 was about 460 GW on average [130] with total generating capacity stated as 786 GW, numbers similar to the present, 2011. At a population of 311 million, this electricity usage is about 1.48 kW per person. (The power usage in the United States scales approximately as the population, which grows at about 1% per year. So an estimate for U.S. power usage in 2020 would be ~500 GW. In China, for example, the growth is faster, 10% per year.) The solar energy influx is roughly 1000 W/m² when the sun is directly overhead. An informed average of incoming solar energy power in the United States over earth rotation is given [131] as 205 W/m². In favorable desert locations (see Figure 1.7b), the average number may be as high as 280 W/m².

10.4.1
Solar Influx, Cell Efficiency, and Size of Solar Field Required to Meet Demand

How large an area would be needed to completely provide the U.S. electricity usage by fields of solar cells at a typical location where the sun's input averages to 205 W/m²? At an assumed 20% conversion efficiency (which is presently the best demonstrated for the CIGS cells, but exceeds present estimates of 12.5% for CdTe modules and 15.8% for CIGS modules; the physics suggests that 20% is possible with these cells), the cell surface area required is 460 GW/[(0.2)(205 W/m²)] = (105.9 km)² = (65.8 mi)². (This number depends on the assumed efficiency and on the solar strength at the assumed location. The GCEP report [131] suggests a 180 km² of land would be required.) So, 100 plots 6.6 miles on edge, or 10 000 plots, 0.66 miles on edge, would generate the total power. (This area, about 0.12% of U.S. land area or ~8% of the area of the State of Iowa, is on the same scale as the area of roads in the United States, 3.6×10^{10} m², equivalent to a square 189.7 km on a side.) Such an installation, implying massive overnight and cloudy day energy storage, which we are neglecting, would solve the U.S. need for electricity on a permanent basis, apart from growth in demand. This approach entirely avoids fuel costs and undesirable gas emissions. There would be no further concern, as regards electricity, for running out of coal or oil, or disposing of spent nuclear fuels.

If thin-film cells to cover a 65.8 mile square were to be printed in a similar fashion as the *Wall Street Journal*, at ~15 m² per copy, with daily circulation of 2 million, the same presses would require $(105.9 \text{ km})^2/(15)(2 \times 10^6) = 374$ days. So a year's production of the *Wall Street Journal* presses would be almost enough to cover the required area with one sheet of paper. This may give some perspective on the scale of production of solar cells that could replace all present sources of electricity in the United States.

10.4 Large-Scale Scenario Based on Thin-Film CdTe or CIGS Cells

From the resource side, for a thin-film cell whose absorber is 3 μm in thickness, the volume of active material required is $(105.9 \text{ km})^2 \times 3 \text{ μm} = 33\,645 \text{ m}^3 \approx 408$ railroad coal cars at 82.5 m^3 per coal car. As a comparison, a trainload, for example, of coal, is around 100 cars. The important point is that we are not talking about a trainload per day, as in shipping of coal, but four trainloads to solve entirely the electricity needs of the country for the future.

An analysis of CdTe cells is similar, with emphasis on the amount of Te required. The density of CdTe is 5850 kg/m^3, with Te comprising 0.53 of the mass, thus 3100 kg/m^3 of Te in CdTe. Again taking the thickness of the absorber layer as 3 μm, with volume 33 645 m^3 as above, the mass of Te needed is 104.3×10^6 kg or 104×10^3 metric tons. (On a volume basis, the Te would be about 0.5 of the volume, thus 16 822 m^3.) This exceeds the annual production of Te, but it appears that new sources of Te are likely to be found.

On another estimate (http://en.wikipedia.org/Wpiki/Cadmium_telluride_photovoltaics), First Solar CdTe modules have area 0.74 m^2 and deliver 73.75 watts (peak). The area of these modules for 460 GW is then 6.2×10^9 m^2 if the sun shines for 24 h, but 1000/205 times larger on the average solar input of 205 W/m^2, thus 3.04×10^{10} m^2. This is the module area, and the land area must be larger to allow access, perhaps to 4×10^{10} m^2. This is a plot of land of side 200 km = 125 mi on a side, and is twice the size of our original estimate. It may be that the CdTe modules are half as efficient as our assumed 20%.

If we scale from the 25 square miles (see Chapter 6) for the 2 GW (peak) solar farm planned for China, the area for 460 GW at average sun energy 205 W/m^2 is $(460/2) \times (1000/205) \times 25$ square miles = 28 050 mi^2, or a plot 167 miles on a side. This is larger than our original estimate of 66 miles on a side and the GCEP estimate of 180 km × 180 km, which is 113 miles on a side.

If we assume First Solar can sell modules at a profit for about $1/Wp, the capital cost of the cells is $\sim (1000/205) \times \$460$ billion. We know that First Solar sold 1.4 GW of modules in 2010. So the problem is of scaling up this production by a factor 328 for 1 year or by a factor of 66 if the production spans 5 years. This deals only with electricity and only in the United States. In the United States, the total average power (all energy sources, coal, hydro, and nuclear) is 3.17 TW, thus 6.9 times the electric power usage we have been describing (see Figures 1.1 and 1.2). Conceivably, all power could eventually be derived from solar PV and wind, converted into electricity, and used for electric heating, electric autos and transport, and so on. In that case, the numbers given above would be multiplied by 6.9 (see Figure 1.7a).

10.4.2
Economics of "Printing Press" CIGS or CdTe Cell Production to Satisfy U.S. Electric Demand

A first generation "printing press" for roll-to-roll processing CIGS CuIn$_{1-x}$Ga$_x$Se$_2$ cells is in operation at the Nanosolar plant in California, which received $150 million in private funding. According to the *New York Times* [129], the company shipped its first solar panels, containing cells printed onto aluminum foil [132] *selling for less than*

$1 per peak watt, in December 2007. (In the rating $1/Wp, Wp is the power that the module would produce when illuminated with the AM -1.5 spectrum at full intensity 1000 W/m^2. The average intensity in the United States is about 205 W/m^2.) The founder of the company was quoted to the effect that his firm was the first to *profitably market cells for less than $1/Wp*. The cells are intended for a 1 MW facility in Germany. It is said that the capacity of the 200 000 square feet facility, where production is based on roll-to-roll processing, is 430 MWp per year. (The company also has plans to produce roof tiles, which have been a popular product of Energy Conversion Devices/ United Solar, a solar energy firm of longer standing.) The efficiency of the cell is embedded in the $/Wp but is not stated by the manufacturer. Upper limit efficiency attained in this class of single-junction cell is 20%. Learning curve effects and even the chance of a tandem version of the mass printed CIGS cell may make 20% a reasonable eventual efficiency. The First Solar CdTe process does not use nano-ink, but nevertheless achieves less than $1/Wp cost. It is also stated that a new CdTe startup, Solextant, will use a nano-ink process (http://Wpww.greentechmedia.com/ articles/read/Thin-FIlm-Solar-Startup-Solexant-and-its-New-CEO/) for CdTe.

Estimates for Nanosolar producing printed CIGS cells depend upon facility cost about $100 million (roughly) for production capacity stated as 430 MWp per year, and cost per watt of the cell (stated as certainly less than $1/Wp in 2007). There is indication that the real cost in successful production may be as low as 30 cents/Wp. According to a rule-of-thumb conversion $1/Wp = 6 c/kWh, so printed CIGS cells (based on our estimate of 30 c/Wp) could be used to market power at a rate 2 c/kWh.

A confirmation of this general scenario is provided by present results for the similar CdTe cells, for which panel cost has been reported as $0.76/Wp, and of which the volume production was 1.4 GW in 2010 (see Figure 10.2).

10.4.3
Projected Total Capital Need, Conditions for Profitable Private Investment

We can use these approximate numbers (a plant to produce 430 MWp per year costs $100 million) to roughly project the capital costs of a possible solar energy future (supply total electricity usage in the United States, neglecting costs of land, storage, and transmission).

If the national average electricity use is 460 GW, and we want to build that capacity over a 10 year period, then building $\approx$ (460 GW/430 MW) (1000/205) (1/10) = 522 plants on the Nanosolar, Inc. model could totally meet that need. This gives an initial facility cost (at $100 million per plant) of $52 billion and a solar cell production cost (at $0.3/Wp) of $0.3 × 460/(0.205 × 10) billion = $67.3 billion per year for 10 years.

Abound Solar (http://Wpww.abound.com/solar-modules/manufacturing) (http:// Www.sustainablebusiness.com/index.cfm/go/news.display/id/21597) has received a $400 million loan guarantee toward expansion of its production capacity from an initial 65 MWp to 845 MWp. From this, we can make an estimate that the capital cost for a production line is about $400 million/800 MWp = 0.5 $/Wp. If we adopt this, and ask how much capital would be required to build capacity in solar modules over 10 years to provide 500 GW continuous usage, the peak capacity will need to be larger by the factor 1000/205 comparing peak insolation to the average value. So, accumulated peak capacity will need to be (500/0.205) GW = 2439 GW. The cost of production lines to build this in 1 year is then $1220 billion, and cost for production lines to build this capacity over 10 years is $122 billion. Again, the production cost for the cells, assuming $0.3/Wp, is $2439 billion × 0.3/10 = $73.2 billion per year for 10 years. Total cost over 10 years is $(122 + 732) billion = $0.845 trillion. If this whole program were paid by the government, it would be $84.5 billion per year for 10 years. The $845 billion is approximately equal to 0.12 of the U.S. military budget ($708 billion in 2010) for a period of 10 years.

Income from this hypothetical investment can be estimated, and indeed exceeds the cost over the 10 year period. The sales value of the produced electricity, whose power ramps up from 0 to 500 GW over 10 years, at $0.1/kWh, is $0.1 × 250 GW × 10 × 365 × 24 × 3600/1000 × 3600 = $ 2.19 trillion. The return on the investment beyond cost is $(2.19 − 0.845) trillion = $1.345 trillion, or 159% over 10 years. On a yearly basis, the return can be calculated as $x^{10} = 1.59$, so $x = 1.047$, or interest rate on annual basis is 4.7%. This is a better return than a savings account, but would be regarded as a very risky investment. And, as we know, banks are reluctant to lend money unless there is a government guarantee, which might be available in the event of a national realization of a state of emergency.

If a national energy emergency should appear, the realization that fossil fuels either are running out or cannot be tolerated for their environmental consequences, leading to a strong government commitment to favor solar energy over fossil fuel energy, such large costs might be undertaken. These costs are several times larger than existing subsidies [133] to energy industries, including oil and coal, but are much smaller than military expenses. The long-term expectation of rising oil prices and recognition of the dangers of coal burning to the climate will probably tilt the economics toward solar and wind generation of electricity. A more benign and likely scenario, suggested by the private investments now occurring in wind energy, is that private capital will be attracted.

A large array of PV (photovoltaics) would normally convert the solar-produced DC (using inverter technology) to conventional AC power on site and connect directly to the power grid. Such a grid connection is conventional or required by statute in many localities for a PV installation, but may be difficult for a large remote installation. We might assume that energy storage, building up from the present capacity in pumped hydropower, will eventually become a larger part of the electric power grid. We will see later that more favorable costs, avoiding a large part of the storage needed, can be found by a nearly equal combination of PV as we have described and wind power (see Chapter 1). Wind power could be built on a scale to cover the night-time demand and PV would provide the added power to supply the peak demand during the day. In this case, the cost of the PV per watt could be reduced almost by a factor 5, since the capacity would be needed only during the middle of the day when the sun power is 1000 W/m². This approach could greatly reduce the need for storage, but would likely require overcapacity in wind power, to cover for a week of cloudy days.

10.5
Comparison of Solar Power versus Wind Power

A simple analysis of the cost of establishing about 500 GW of continuous power from photovoltaics indicates that a solar farm around 150 miles on a side and costs (neglecting the land) for the production lines and cell production of about $ 0.845 trillion would be needed. In Chapter 1, a similar analysis for wind turbines required a land area about 230 miles on a side and a capital cost about $ 0.5 trillion. We noted there that the cost would be similar to 7% of the U.S. military budget over a period of 10 years. In neither case were the costs associated with storage and transmission of energy taken into account. In both cases, storage and transmission of energy could be envisioned by production of hydrogen from water, with the hydrogen used as the portable form of energy. (The primary form of storage in the U.S. power grid is pumped hydroelectric power, with capacity about 21 GW at present (http://www.esd.ornl.gov/WindWaterPower/PSHSummit.pdf) with worldwide capacity 127 GW.)

The great difficulties with these energy sources are the diffuse nature of the energy available, requiring extremely large collection areas, and the variability of the resulting power output. The advantage of the wind turbines is that the wind keeps blowing at night, while solar power is unavailable. The basic daily variation of the photovoltaic power output can be used to advantage, however, if the renewable power is derived from nearly equal amounts of wind and solar energy. The variation in

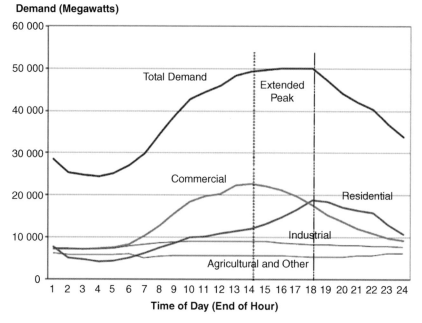

Figure 10.5 The total load profile (http://www.mpoweruk.com/electricity_demand.htm (Data from Lawrence Berkeley National Laboratory, U.S. Dept. of Energy)) for the State of California on a hot day in 1999. The demand during the day is about twice the demand at night.

power demand over 24 h for the State of California on a hot day is illustrated in Figure 10.5. (The pattern will differ on a weekend or on a cold day when air conditioning is not needed.) But, approximately, the electric power demand is doubled during the day over that at night. So, a large portion of the variation could be accommodated without storage by equal amounts of photovoltaic and wind electric power.

The combined renewable power scenario for average demand 500 GW to be provided solely by equal amounts of solar and wind power would cost $250 billion (for wind) plus $845 billion × 0.5 (205/1000) = $86.6 billion for solar. The cost of the total on this estimate is $337 billion.

To be more realistic, we might double these numbers to allow for a period of cloudy weather, for example. The extra capacity on good (sunny and windy) days would be an opportunity to generate hydrogen, which might eventually fuel automobiles, and otherwise has a market value. If we ask about the total energy usage in the United States we multiply this number by 6.9.

10.6
The Importance of Storage and Grid Management to Large-Scale Utilization

In spite of the comment above, storage capacity is an essential need for large use of either solar or wind energy. The primary storage need is with solar, because the sun is absent at night. Storage is also a (reduced) problem with wind power, since the wind blows more strongly at night, and weather is variable. Clouds greatly reduce the power output from a photovoltaic facility. The question of variability in renewable energy sources is being addressed in China, which has unquestionably decided it needs as much power as it can get, with strong planned growth in hydropower (demonstrated by the Three Gorges Dam hydropower plant) and solar and wind farms. China's State Grid is the largest power grid in the world and is expanding. Its President, Liu Zhenya, is quoted as saying that the multiple "Three Gorges of Wind" that are to be built in China will have to be "bundled" with natural gas, coal, and nuclear [134]. This does not mention storage, but it is likely that there is considerable pumped hydro storage in China (Figure 1.11).

Hydrogen storage is sensible as a means of storage for both wind and solar electric power. This is not presently being done. Hydrogen storage can be done locally on almost any scale, from the personal or family house scale addressed by Nocera (see Section 9.2) to large scales as appropriate for wind or solar farms. The grid-scale storage that is most common is pumped hydroelectric capacity currently at 27 GW in the United States (see Figure 1.11.) The use of hydrogen as an energy intermediary, including its use with fuel cells to power electric cars, has been discussed in Chapter 9.

The PV-based hydrogen energy storage approach can also be envisioned on a small scale, in the vision of Nocera discussed in Chapter 9, similar to Figure 10.7. Nocera is working with (http://uk.ibtimes.com/articles/20110328/leaf-mimicking-catalyst-inventor-signs-huge-contract-with-tata-group-hydrogen-from-water-device.htm) the

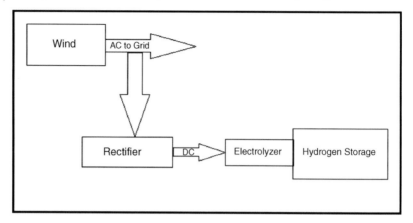

Figure 10.6 Sketch of off-the-shelf equipment for creating hydrogen from water with unused wind electric power.

Tata group in India to make a unit suitable for individual houses to provide electricity with energy storage in the form of hydrogen. The hope is to market many small units to villagers in rural India, providing power and mitigating, to some extent, the rapid construction of coal burning power plants (said to be four per week in India and China) with their known adverse effect on the climate [135].

Storage as touched on in a discussion of a planned European grid of wind power sources is given in *The Economist* [136]. This article notes an advantage of linking Norway to a proposed European wind power grid: to gain storage capacity. It is stated that the energy storage capacity obtained by pumping water uphill into storage above existing hydroelectric dams in Norway is sufficient to completely power the proposed European wind power network over a period of 4 weeks. If this statement is correct,

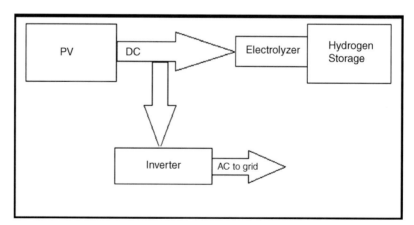

Figure 10.7 Sketch of off-the-shelf equipment for creating hydrogen by electrolyzing water with unused photovoltaic electric power.

then a similar opportunity likely exists for energy storage in hydroelectric facilities in the United States and Canada. Installed hydroelectric capacities are listed [137] as 27.53 GW in Norway, usage 0.49, compared to U.S. capacity 79.5 GW, usage 0.42, and Canadian capacity 89.0 GW, usage 0.59 [137]. Present pumped hydro capacity in the United States is 27 GW.

This suggests that the typically unused U.S. hydro capacity, about 46.1 GW, with installation of uphill pumping capacity, could provide about 10% storage capacity for all U.S. electric power.

10.6.1
Batteries: from Lead–Acid to Lithium to Sodium Sulfur

Beyond hydroelectric and hydrogen storage (presently nearly hypothetical), a practical approach to storage is in large-scale batteries [138].

A large-scale battery project undertaken by a U.S. utility entails "6 MW" of capacity at a total cost of $27 million, thus a capital cost of $4.5/Wp or $0.64/Wh. The basic units, rated at "1 MW," are about the size of a double decker bus and use sodium–sulfur chemistry and operate at about 800 °F. The battery is said to deliver 1 MW for about 7 h, so a more conventional rating for the capacity would be 7 MWh per battery. The batteries are said to be 80% efficient, and will be used to smooth power from wind turbines and make that power easier to connect to their basic coal-fired generating electrical grid. The batteries [138] will "charge at night, when the wind is strong but prices are low, and give the electricity back the next afternoon, when there is hardly any wind but power prices are many times higher." A sodium–sulfur, NaS, battery of 4 MWp, total 32 MWh capacity and cost $25 M, is reported (http://inhabitat.com/bob-americas-biggest-sodium-sulfur-battery-powers-a-texas-town/) in Presidio, Texas.

This cost is then $0.78/Wh of storage. It appears that these stated costs are higher than necessary, compared to $400/kWh = $0.4/Wh given by U.S. Department of Energy for NaS batteries [139]. The same article estimates the cost of pumped hydroelectric power as 0.1$/Wh.

Rechargeable batteries reversibly interchange chemical and electric energy. These are categorized by open-circuit voltage (V), and electrical energy (Wh/kg) or (Wh/l). The equivalent circuit is an electromotive force voltage, EMF, arising in the underlying chemical reaction, with a series resistance (Figure 10.8).

A battery consists of an array of individual cells that are connected in series to obtain a desired total voltage and in parallel to reach a necessary current capacity. Often, the peak rate of discharge and maximum power output are important.

The traditional lead–acid automobile battery consists of dense and heavy materials, lead and sulfuric acid, and is used primarily in high current applications such as starting conventional automobile engines.

Energy density is at a premium in modern applications, for example, in cellular telephones, and also in larger scale applications such as hybrid automobiles. Li-ion batteries are used in consumer electronics, notably cellular telephones and laptop computers, because of their high energy density, in the range 160 Wh/kg and 350 Wh/l. The Li-ion battery has the largest portion of the portable battery market,

218 | 10 Large-Scale Fabrication, Learning Curves, and Economics Including Storage

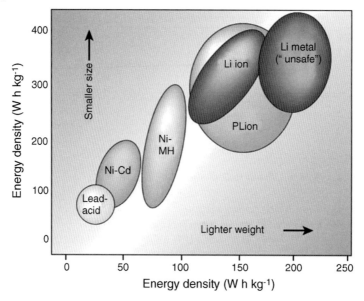

Figure 10.8 Comparison [140] of energy densities for several types of batteries. Units are watt hours per liter (ordinate) and watt hours per kg (abscissa). Here, the label "PLiON" refers to a class of rechargeable plastic Li-ion batteries.

about 63%, mostly in electronics, but not yet in hybrid cars. Ni–Cd batteries are used in power tools.

The Prius hybrid car uses NiMH batteries, presently considered a more conservative choice than the Li-ion battery, for which as shown in Figure 10.9 the graphite electrode may present a fire hazard. Nonetheless, an expensive roadster is being sold by Tesla that is purely electric and powered by 6831 Li-ion cells of the type used in laptop computers. This vehicle accelerates to 60 mph in less than 4 s and has a range of 210 miles [141].

10.6.2
Basics of Lithium Batteries

A diagram of a lithium ion battery is shown in Figure 10.9.

Variations on the basic lithium battery shown in Figure 10.9 include a cathode of iron phosphate LiFePO$_4$ [143] (manufactured by Lithium Technology Corp. and by A123Sysems, which may involve nanoparticles) and possibly an anode consisting of lithium titanate nanoparticles (Altairnano, Inc.) that will not burn.

It appears [143] that the electrical conductivity of the iron phosphate LiFePO$_4$ cathode (LFP), which had been low, was increased by doping with metals such as aluminum, niobium, and zirconium, and also probably involving nanoscopic carbon particles. These advances now make the iron phosphate cathode workable, to allow a higher discharge current, fast charging time, and stability under extreme conditions. The basic advantages of LFP cathodes over the cobalt cathodes are low cost, high

(a)
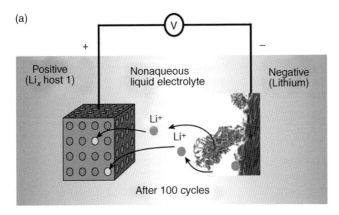

(b)
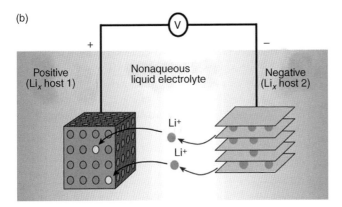

Figure 10.9 Operating principle [142] of rechargeable Li-ion battery. Li ions diffuse from high-energy sites in the graphite anode to low-energy sites in the cathode, driving charge around the external circuit. Cells using $Li_{1-x}CoO_2$ (*left*) and Li_x intercalated into graphite (*right*) provide 3.6 V, energy densities 120–150 Wh/kg are widely used in portable electronic devices. Li ions reversibly enter (intercalate) and leave weakly bound positions between the graphene carbon layers of graphite.

abundance of iron, and freedom from overheating. Another advantage is that in the chemical reaction in the LFP case, relative to that shown in Equation 10.3, the value of x goes completely to zero, leaving no residual Li in the cathode when fully charged.

The chemical reaction for one form of Li-ion cell is

$$Li_{1-x}CoO_2 + Li_xC_6 = C_6 + LiCoO_2, \tag{10.3}$$

where the carbon is a graphite electrode. *Ions, not electrons, are the current carriers.* Note that lithium ions are not oxidized. In a lithium ion battery, the positive lithium ions flow internally from the graphite anode to the cathode, with the transition metal, cobalt, in $Li_{1-x}CoO_2$ being reduced from Co^{4+} to Co^{3+} during discharge. The performance may be as high as 160 Wh/kg.

The performance of Li-ion batteries can be improved by nanostructuring the electrodes, to enlarge the effective surface area. For example, it is stated that both higher power and higher storage capacity can be obtained by applying the active anode and cathode materials in a very thin film to copper nanorods anchored to sheets of copper foil. Enlarged electrode surface area by nanostructuring can increase the short-circuit current of a battery, assuming good electrical conduction to the nanostructured areas.

10.6.3
NiMH

A nickel metal hydride battery, abbreviated NiMH, is a type of rechargeable battery, with 1.2 V nominal voltage, similar to a nickel–cadmium battery, but has an anode of hydrogen-absorbing alloy, instead of cadmium. The anode reaction occurring in a NiMH battery is

$$H_2O + M + e^- \leftrightarrow OH^- + MH \tag{10.4}$$

The battery is charged in the forward direction of this equation and discharged in the reverse direction. Nickel (II) hydroxide forms the cathode. The metal M in the anode of a NiMH battery is typically an intermetallic compound. Several different intermetallic compounds have been developed, which mainly fall into two classes. The most common form of anode metal M is AB_5, where A is a rare earth mixture of lanthanum, cerium, neodymium, and praseodymium and B is nickel, cobalt, manganese, and/or aluminum. Other batteries use higher capacity negative electrodes based on AB_2 compounds, where A is titanium and/or vanadium and B is zirconium or nickel, modified with chromium, cobalt, iron, and/or manganese. Any of these compounds M serves the same purpose, reversibly forming a mixture of metal hydride compounds. When hydrogen ions are forced out of the potassium hydroxide electrolyte solution by the charging voltage, it is essential that hydride formation is more favorable than forming a gas, allowing a low pressure and volume to be maintained. As the battery is discharged, these same ions are released to participate in the reverse reaction.

NiMH batteries have an alkaline electrolyte, usually potassium hydroxide. A NiMH battery can have two–three times the capacity of an equivalent size NiCd battery. However, compared to the lithium ion battery, the volumetric energy density is lower. The specific energy density for the NiMH battery is approximately 70 Wh/kg. It is reported that a NiMH Prius car battery contains 12 kg of lanthanum. An advantage of the NiCd battery over the NiMH is typically higher short-circuit current.

Batteries used in fully electric vehicles EV (Nissan Leaf) and plug-in electric vehicles PHEV (Chevy Volt) are in the capacity range 20–50 kWh. The batteries in hybrid electric vehicles, HEV, like the Toyota Prius are generally in the range of 2 kWh.

The broader question of energy storage in the power grid can be discussed in terms of many small batteries. The total sales of the Toyota Prius are about 1 million, half in the United States. The storage capacity of 0.5 million Prius cars, assuming 2 kWh per car, is seen to be 2 GWh. Compared to Figure 10.5 for the State of California, the peak power is 50 GW, and it is seen that the application of all the Prius batteries in the United States to the grid in California at this peak power would not be appreciable, it would meet demand for only $2/50\,h = 2.4$ min.

If the cars were of the EV or PHEV category, assuming 50 kWh battery capacity, and if the number of cars is 0.8 of the population of Calfornia (37 million), thus 29.6 million, then the storage capacity would be $29.6 \times 10^6 \times 50\,kWh = 1480\,GWh$, which is enough to supply peak demand in Figure 10.5 for about 3 h.

One estimate of the cost of this storage capacity could be based on $30 000 per PHEV car, which would give $0.89 trillion.

The cost of 1480 GWh of sodium–sulfur battery storage at $0.4/Wh is $0.592 trillion and would be $0.148 trillion if provided by pumped hydro storage. These numbers exceed the State of California budget, on the order of $100 million. It appears that the cost of storage, illustrated here as extremely high, is larger than the cost of generating capacity. For the case of California at peak demand in Figure 10.5 as 50 GW, the nominal cost of that electric generating capacity at $1/Wp would be $50 billion. Reducing the cost of battery storage is a goal of ongoing research and development in the U.S. Department of Energy.

Larger Li-based batteries have been described for buses and vans [144]. The batteries described are 85 kWh, and are based on lithium iron phosphate technology.

A possible approach to the electric energy storage might be "one battery per person." The average power per person in the United States is 1.48 kW, as we saw at the beginning of this chapter. (This figure includes industrial and other electricity uses and is higher than the needs of the typical individual.) So an individual owning one of the 85 kWh batteries would have a buffer $85\,kWh/1.48\,kW = 57.4\,h = 2.4$ days in case of a power outage. In urban areas, it is common practice for each apartment building to have a water storage tank on its roof. Adding a set of batteries to the roof would not be so different. If the price per watt hour is $0.40/Wh, then the 85 kWh battery cost would be $34 000. This is very expensive. It may be more economical to buy an auxiliary generator. An apartment building with 100 persons might need 30 kW, which can be purchased for about $8000.

The lowest cost form of energy storage does seem to be pumped hydro power, which, however, is a large-scale grid solution and does not help the individual or isolated wind or solar producer not connected to the power grid. Pumped hydro is available only in select locations, many of which are already developed.

By way of summary, enough solar-based energy is available (see Figure 1.7a), but means of harvesting the energy (to create electricity or a portable fuel like hydrogen, methane, or gasoline) are lacking. Sources of energy include direct sunlight, and its secondary influences, which are winds, hydroelectric power, and the energy of water waves. The sunlight falling on an area about equivalent to the area of the existing U.S. highways could be converted, using solar cells, into electricity to replace entirely the power plants now in use. A rather similar situation exists with wind power. A significant fraction, 20%, of electricity in Denmark comes from wind turbines. The primary common problem with solar and wind energy is the intermittency of the power. This can be solved either with storage or by incorporating many locations into a common grid, with the idea that when one location is quiet another will be strongly contributing. The second major problem is that the best locations for collecting solar and wind energy are often remote from population centers, with a large need for transmission of power. A significant step to reduce the need for storage, as we have suggested above, is to combine approximately equal amounts of wind and solar energy, with the solar energy providing the extra power needed during the daytime. An extensive plan was produced for continental Europe and North Africa [145], concluding that with large use of DC power transmission lines and undersea cables, which are more efficient at longer transmission distances, the whole area could be serviced with renewable energy, without need for oil and coal, and at costs similar to present costs. An optimization resulted in clustering renewable wind- and sun-based sources where the conditions are optimum, notably in North Africa, and sending the energy to population centers with extensive use of DC power transmission lines. In the United States, this would suggest putting most of the solar cells in the far southwest, see Figure 1.7b. The optimization favored wind over solar on a cost basis, and in the category of solar favored solar thermal, known also as concentrated solar power (see Section 5.3) over solar cells on a cost basis and because of its option of heat energy storage, based on molten salts. The optimization also included Norway, despite the expense of undersea transmission cables, because of its storage capacity.

In this chapter, we have examined, from a simplified technical point of view, approaches to replacing U.S. electric production entirely by renewable technology, emphasizing photovoltaic cells, with crude estimates on costs and comments on storage and grid management. In Chapter 11, we will venture some estimates of what may actually happen, taking into account external factors that include the world's reaction to the growing recognition of climate change as associated with carbon-based energy, and the variability among countries in the efficiency of their political processes toward recognizing and acting on needs for change.

11
Prospects for Solar and Renewable Power

The usage of renewable power is small at present, but it is growing rapidly. We saw in Chapter 1 that of the 14.7 TW world total power consumption in 2009, the largest renewable contribution was hydroelectric power at about 7.3%. The contributions of wind and solar power were too small to be seen on the overall linear plot of Figure 1.1. In Figure 1.2 (for the United States), out of a total renewable portion of 8%, only 1% was solar and 9% was wind. Solar and wind account for less than 0.8% of U.S. power.

11.1
Rapid Growth in Solar and Wind Power

The present growth rates are extremely high. For example, solar electric installed capacity in the United States is said to have increased at 74% per year since 2008 [146]. Investment overall in renewable energy, according to the Bloomberg New Energy Finance report of 2011 ([146b], Figure 3), has increased at an average rate of 36% per year in the period 2004–2010, with global investment in 2010 of $211 billion. (To calibrate the size of this investment, taking the world's energy consumption at 15 TW, a rough estimate of the value of the production capacity, at $1/W, is on the order of $15 trillion, 71 times larger.)

Projections of these rates suggest a quickly changing global energy mix. Assuming the renewable capacity grows at the same rate, 36% per year, as the investment, and neglecting growth in nonrenewable energy, one can ask how long it would take to reach 50 from 1%. The answer is 12.7 years, but this is a very uncertain extrapolation to be sure.

A more cautious limit on the ultimate contribution of *photovoltaic* capacity, as 15% of worldwide electric power by 2050, has been offered by Maycock, a seasoned participant and observer of the photovoltaic industry, as quoted by Yergin [147].

Mr. Maycock comments: "15% of the world's electricity is a very big number. To reach 15% will require trillions of dollars of investment. For a business that is now doing sixty billion dollars a year, that is a very nice mountain to be challenged by."

This comment implies a lower growth rate: to go from 1% (roughly) to 15% in 38 years corresponds to a growth rate of 7.4% per year. The recent enormous growth

Nanophysics of Solar and Renewable Energy, First Edition. Edward L. Wolf.
© 2012 Wiley-VCH Verlag GmbH & Co. KGaA. Published 2012 by Wiley-VCH Verlag GmbH & Co. KGaA.

rate, 36% per year, extending over 6 years for renewable energy, is likely to slow down when the market share becomes larger. Still, we will argue that there is only one direction, upward, for renewable energy capacity, firms, and workers. This is a booming (small) business that has a long period of growth in store. This of course does not mean that every start-up company will succeed; there is already fierce competition in the area of solar panels.

We give more details on the recent growth rates of the renewable categories of solar and wind capacity (which at present total to less than 1% of worldwide total energy production). Solar *photovoltaic* global production doubled in 2010 [148], according to the report REN21 (p. 12). This report indicates that in 2010, 17 GW of capacity, representing 100% growth over the previous year's growth, was added bringing total PV capacity to 40 GW. (This global capacity is about 5% of U.S. electric generating capacity, about 800 GW.) Figure 11.1 gives details of investment in 2010 in several categories of renewable energy.

The investments indicated in Figure 11.1 total to $211 billion for 2010, and indicate a preponderance of activity in wind and solar technology, with declining investment in some other renewable categories. The category "solar" is composed primarily of solar photovoltaic and solar thermal (concentrated solar power), and apparently does not include solar water heating. It appears that the growth for "solar" of 52% in Figure 11.1, compared to 100% for "photovoltaics" (above), indicates that the growth rate for concentrated solar thermal power (CSP), which was discussed in Chapter 5, while large, lagged the PV sector in year 2010.

Solar water heating and home heating is not included, although installed capacity was recently approximately 10 times larger than for solar electrics. These systems replace electric power from the grid, and according to Richter [149], a recent value of

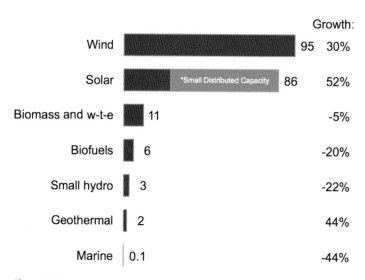

Figure 11.1 New investments in renewable energy in 2010 by technology, units are $bn and growth is relative to previous year 2009, according to Bloomberg New Energy Finance [146], Figure 6, p. 14.

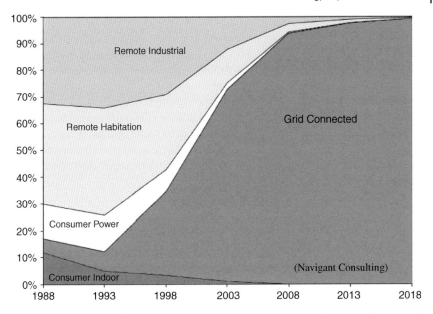

Figure 11.2 Trend toward grid-connected photovoltaic installations [150]. The total capacity has been increasing strongly, so this figure does not imply a reduction in capacity in "remote industrial" or "remote habitation" categories, but rather that the "grid connected" category, which includes large "solar farm" installations plus grid-connected rooftop installations, has been rapidly increasing and is projected to further increase.

the equivalent power capacity was 105 GW, much larger than the solar electric capacity at that time. It is reported that 20 GW of solar water heating capacity was installed in 2007 [148]. These systems have no problems either with intermittency or with distribution of the energy. For heating water, such a system is 10 times more efficient than using solar electric power to run a water heater.

In Figure 11.1 the "solar" category shows about two-thirds of the investment as "small distributed capacity," compared to large installations. This is described as "overwhelmingly photovoltaics on rooftops," which have been subsidized especially in Germany and are well known in California. (A scenario of rooftop photovoltaics proposed for New York City was described in Chapter 5.) The rooftop photovoltaics are predominantly grid connected, and illustrate a trend noted in Figure 11.2.

11.2
Renewable Energy Beyond Solar and Wind

Figure 11.1 indicated major but falling investments in biomass and biofuels in 2010 versus 2009. Also noted are investments in "small hydropower" and marine (water turbines and wave energy devices), and in geothermal. Several of the energy topics we have classified as "renewable" and covered in this book are not listed in Figure 11.1. This might mean the numbers are too small to list or even that no activity is occurring, or, more likely, that the item is not recognized as "renewable" energy in the financial world. We know there is major investment in nuclear fusion that was covered in Chapter 4, but this area is not traditionally recognized as renewable energy. To be sure, the ongoing work in Tokamak reactors is purely research, since there is no product. Investment in battery technology and capacity is known to be taking place in connection with the hybrid and plug-in electric automobiles. Hydrogen storage is not evident in these listings, but we are quite sure that the efforts of Nocera (Section 9.2) in collaboration with the Tata group in India have value and are going forward. Whether the "artificial leaf" idea that was described in Chapter 9 will be commercially successful is not clear, but it has great potential for growth, especially in developing countries and in the vast third world.

11.3
The Legacy World, Developing Countries, and the Third World

The electric grid is a feature of the *legacy world* (whose citizens inherit a high standard of living) and of the developing countries, but it is not a feature of the third world. Poverty, in the global context of 7 billion people, is defined as earning less than $1.25 per day, and it is said that there are 2 billion small farmers. Some of the renewable energy approaches, perhaps comparable to the cell phone, for which there are 4.6 billion subscriptions, can be deployed in the third world. Solar cells, solar water heaters, and the "artificial leaf" energy conversion system can be deployed on a small scale, on a rooftop. It has been suggested that renewable energy can put an end to "energy poverty": that electric power, even in small amounts, will allow persons to make contact with the outer world of ideas. This can start a path to prosperity [151] for people in the third world. It is noted that 1.4 billion people are not on an electric grid, and are effectively cut off from the rest of the world. From the point of view of renewable energy workers, this is both a great opportunity and a challenge.

The challenge of providing energy, in coming decades, for even the developing world is severe. Figure 11.3 plots the numbers of cars per 1000 persons in legacy countries (bottom three in Figure 11.3, total population ~1 billion) and developing nations (top three, total population ~2.5 billion). The population of Africa, not represented in Figure 11.3, is about 1 billion.

The suggestion is that the present growth patterns in developing countries may lead to a world with more than twice as many autos and thus, even with growing efficiency, more than twice the need for fuel. Indeed, it is reported that in 2010 the number of cars in the world reached 1.015 billion (http://www.huffingtonpost.ca/2011/08/23/car-population_n_934291.html), with a growth rate of 3.6%, which implies doubling the number of cars in 20 years (http://e360.yale.edu/content/feature.msp?id=2128). The growth in the number of cars in China in 2010 specifically was 27.5%.

It is already the case that auto sales in China exceed those in the United States. It is estimated that 2 billion autos would require oil production of 120 million barrels/day

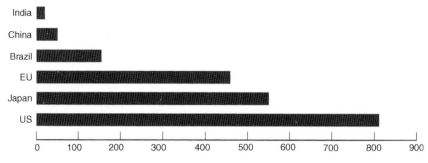

Figure 11.3 Number of cars per 1000 population in developing countries (*top*) and the legacy world (*bottom*) in 2010 [147]. The number of people in the developing countries listed is about 2.5 times the combined population of the EU, Japan, and United States. Rapid growth seems assured in these developing countries, assuring rapid increase in energy demand.

(http://www.huffingtonpost.ca/2011/08/23/car-population_n_934291.html), compared to the present value of 92 million bbl/day [152].

A similar pattern is likely to exist for electricity and total energy use. It has been predicted [153] that over the next 20 years the growth in China's economy will average 7% per year, compared to 11% over the past decade. This is consistent with a statement that electric power demand in China is growing at a 10% rate.

11.4
Can Energy Supply Meet Demand in the Longer Future?

This question is addressed at length in the recent book of Daniel Yergin ([147], p. 704). We have seen in Figure 1.1 that oil is the largest single source of energy, providing about 6 TW, or about 40% of the total global power consumption. Oil is also the most threatened energy source, well known to be limited in supply, estimated as 1.4×10^{12} barrels, compared to 2.4×10^{12} barrels [152] in 1900. This implies that 42% of available oil has been extracted since 1900. The remaining accessible oil reserve can be expressed as the number of years at a production level of 100 million bbl/day, and this number is 38.4 years, taking Yergin's reserve figure. (The present production is 92 million bbl/day, and rising.)

11.4.1
The "Oil Bubble"

Informed views have been expressed [147, 152] that exploration and more extreme techniques of extraction will keep oil production rising modestly for a few decades. The historical plot of oil production [152] shows a roughly linear rise from around 15 million barrels per day in 1950 to 92 million bbl/day in 2011. Optimism is expressed in that the rate of extraction is still rising. In fact, production has grown by

30%, in an accurately linear fashion since 1978. Yergin specifically predicts that production will *continue to rise for 20 years leading to a production of 110 million bbl/d in 2030*. Adhering to Yergin's figure of 1.4 trillion barrels of "accessible oil" remaining, the rising production implies that the remaining 58% will be used more quickly. The production curve [152] if extended linearly, as Yergin predicts, from 92 million barrels per day in 2011 to 110 million bbl/day in 2030 and beyond allows a simple estimate of about 35.5 years, to expend the remaining 58% of the oil. On this model, the last of the "accessible oil" [152] will be used in 2046.

A similar prediction is made by Richter [154], who says that the next 30 years will expend as much oil as have been used to date, the latter about 10^{12} barrels. He points out, however, that there will continue to be "heavy oil" at about three times the price, coming from tar sands, bitumen, and shale. Richter says, "Oil will start to become much more expensive after 2030. . ... Other methods of fueling transportation will be needed."

Since oil production before 1950 accounts for very little of the total oil, the extrapolated production curve, for accessible oil, following Yergin's prediction, can be described as an "oil bubble" extending from about 1950 to 2046. The details of the extrapolated production curve past 2030 may vary, but the era of inexpensive energy will certainly end. (It is also predicted by oil executives, citing increased demand from China and India, that the price will rise significantly above the recent peak level of $147 per barrel, apart from any question of falling supply [155].)

A sharply rising price will lead markets to shift to other energy sources, such as gasoline refined from natural gas and coal, and to various renewable sources of energy. The African firm Sasol has produced 1.5 billion gallons of gasoline (http://www.slate.com/articles/business/moneybox/2006/10/thanks_for_the_cheap_gas_mr_hitler.html?nav=ais) from coal and gas, based on the Fischer–Tropsch chemical process. The supply of coal has been estimated as enough to last until 2080 [154]. The rising price of oil will be good news for sellers of renewable energy, although it will be disruptive to the world economy in many ways. It will have its greatest effect on the legacy world, notably the owners of the 2 billion autos, but will have almost no impact on the third world.

Adapting to disruption will be easiest in the developing countries, because they are able to build new facilities as appropriate, rather than face a task of modifying or replacing old facilities. China in particular has a strong central government that seems to be forward-looking and able to take executive steps without hesitation to adapt to new situations. It is building many new coal-fired electric plants, by necessity, as well as researching ways to bury the carbon dioxide rather than to release it into the atmosphere. China is going rapidly ahead with renewable energy in hydroelectric power, as exemplified by the Three Gorges dam, in "Three Gorges of Wind" farms, and in solar electric facilities. It is taking steps toward electric cars. China recognizes the importance of an electric grid to integrate the renewable sources. Since the cost of energy storage is so high, the problem of intermittency in wind and solar energy seems best solved by combining sources in a large network [156].

While innovation often does not originate in China, Chinese firms are taking over the manufacture of the new items, especially solar panels and the blades for wind farms. We saw in Chapter 10 that China has taken 65% of the solar panel market and has driven the solar panel price down to $1.20/Wp. It appears that wind turbine manufacture is likely to remain in the United States and Europe. China will be a strong force in adapting to the new energy world for its own citizens and also as a business opportunity for export.

In the legacy world, planning and execution of needed changes may increasingly originate in large cities that are economically viable and not controlled to such a large extent by vested interests opposed to change. The European Union has taken strong steps toward integrating solar and wind power into its electric grid. It seems likely that bottom-up efforts like rooftop solar cells and the "artificial leaf" technology may become important, especially in the third world of several billions. While the legacy energy industries, especially involving the electricity grid, are slow to change, a bottom-up technology that does not need government approval, can appear almost overnight, as happened with cellular telephones. It is hard to predict what may happen. Perhaps, a Steve Jobs counterpart will arise in the energy area. The late Steve Jobs, CEO of Apple, was a genius in identifying new combinations of existing technology to meet needs and indeed to create new markets. As we have noted, the energy market is the largest of all possible markets.

Following recent data and predictions [152], we have rationally predicted that the "oil bubble" for conventional accessible oil will extend to about 2046. There will still be oil for those who can pay for it, but the mass market will no longer be dominated by oil. The world will not be without energy, because there is enough coal to last for a longer time, perhaps until 2080 [154]. Still, there is a high level of concern on the adequacy of energy supplies, especially in the face of growth in the developing world, and in view of the climate changing aspects of fossil fuel use.

11.4.2
The "Energy Miracle"

Bill Gates, the former CEO of Microsoft, is now a well-informed practicing global philanthropist. In an important address, he explained why, if he could have one wish to improve the lot of humanity over the next 50 years, he would choose an "energy miracle" [157].

The "energy miracle" would be a new technology that would produce energy at *half the price and with no carbon dioxide emissions*. The reasoning is that a low price for the new technology is the only way that developing countries, notably China and India, can avoid greatly increasing the use of coal to meet their energy needs, with increased carbon dioxide emissions. Thus, carbon dioxide emissions, leading to climate change, are clearly identified by Mr. Gates as one of the greatest dangers to humanity on a 50 year timescale.

The science of climate change is beyond the scope of this book. The danger of greenhouse gas accumulation, leading to climate warming, is scientifically accepted. The U.S. Supreme Court in April 2007 ruled that CO_2 was a pollutant that "may be reasonably anticipated to endanger public health and welfare" and that the Environmental Protection Agency, if it failed to regulate CO_2 emissions, would be "capricious" and "not in accordance with the law" ([147], pp. 502–503). The risk and harm that might accrue (from nonregulation) might include costly storms and loss of shoreline, risks that were "both actual and imminent."

The cautionary point is that Gates identifies the desired technology as a "miracle." In this book, we have examined quite closely many of the leading possible sources of energy to replace the burning of coal, gas, and oil. Can we see among these any technology that will provide energy at half the cost with no carbon dioxide emissions? The answer is, probably not.

But the good news is that any or all of the renewable technologies that we have discussed that can approach \$1/W in capacity cost will be in increasing demand as the costs of fossil fuels go up and the damage that they do to society at large becomes more widely recognized and compensated. Perhaps, it will be the good fortune of a new Steve Jobs to identify and manufacture a winning combination of these renewable technologies.

This book contains a lot of material on the sun, how it generates energy, and on attempts to make the same nuclear fusion process work on earth. Unfortunately, these technologies do not at present seem to be leading prospects for the energy shortfall that is foreseen. There is some chance that innovation in the Tokamak-type fusion reactor may occur, and the time available is 20–30 years. There is also some chance that nuclear fusion reactors based on thorium, rather than uranium, may advance. The advantages of energy based on nuclear reactions are their high energy density and the continuous supply of power that is available.

Energy security was famously described by Winston Churchill as arising from "variety, and variety alone" referring to sources of oil as he planned to shift the British fleet from coal to oil power. The same dictum may apply to the diffuse sources of solar and renewable energy that will become more important after the "oil bubble" has passed, approximately 2046. Solar thermal, solar photovoltaic, solar water heating, wind, hydroelectric, and wave energies will all be available, and all may be needed to fill the shortfall that will appear, accompanied by rising demand from developing nations. As the price of oil rises approaching 2030, a wider variety of cars will likely appear, evolving from today's mix of hybrid, plug-in hybrid, and full electrics. Whether the long-range nonpetrol auto of the future mass market will run on better batteries or on a tank of a fuel such as hydrogen via fuel-cell electric propulsion seems to be an open question. In any case, it seems the renewable energy business will be booming.

Appendix A: Exercises

Exercises to Chapter 1

1.1 The power density from the sun on a clear day is about 1000 W/m². It is stated that the average electric power consumption in the United States was 460 GW in 2004. How large an area would be needed to get all this energy from sunlight, assuming the collection works 30% of the time (no sun at night) and at 10% efficiency. Express the area as miles × miles.

1.2 If the average power usage in the United States in 2004 was 460 GW, and there were 294 million people, what is the average power per person? Why is this so high?

1.3 The energy of a single photon is expressed as $E = hf$, where h is Planck's constant 6.626×10^{-34} Joule second and f is the frequency of the light in Hertz. The sun's spectrum peaks near wavelength $\lambda = 486$ nm. Assuming 486 nm, using the relation $c = f\lambda$, where $c = 3 \times 10^8$ m/s is the speed of light, what is the energy per photon? How many photons per second fall on 1 m² in full sunlight? If the observing area is approximated as the area of a single atom, approximated as $A = \pi a_o^2$, where $a_o = 0.052$ nm is the Bohr radius of the hydrogen atom, how many photons per second fall on a single hydrogen atom?

1.4 The radiation outward from the earth is estimated in Chapter 1. Find the percentage by which this radiated power will change if the earth's temperature rises by 2 K from an assumed value of 288 K.

1.5 The consumption of oil in the United States is 92 million barrels per day. If a barrel (bbl) of oil is 0.159 m³ and costs $100, find the volume per year in m³ and in acre-feet and find the value in dollars. (One acre-foot is 1233.5 m³).

1.6 Using the relation shown just after Equation 1.2, $\lambda_m T = \text{constant} = 2.9$ mm K, find the wavelength at the maximum of the Planck distribution function for $T = 288$ K, approximately the temperature of the earth. Find the energy of the photon at this wavelength, in eV.

1.7 The energy content in a gallon of gasoline is approximately 125 MJ. If a car is rated at 30 miles per gallon, how many Joules per mile does this correspond to? What is the cost per mile for this car?

Nanophysics of Solar and Renewable Energy, First Edition. Edward L. Wolf.
© 2012 Wiley-VCH Verlag GmbH & Co. KGaA. Published 2012 by Wiley-VCH Verlag GmbH & Co. KGaA.

1.8 The Chevy Volt is said to use 0.36 kWh/mi in full electric mode. How many Joules per mile does this correspond to? If the cost of electricity is 10 c/kWh, what is the cost per mile for this car?

1.9 If gasoline costs $4/gallon, and the energy content is 125 MJ per gallon, what is the energy cost expressed in $/kWh?

1.10 According to *The Economist*, Oct. 15, 2011, the Agua Caliente solar farm in Arizona is rated at 290 MW and will cover 1750 acres when completed in 2014. The plant will have 5 million CdTe solar panels. What is the power per panel? If the sun's intensity is 1000 W/m^2, what is the efficiency of this plant? One acre is 4050 m^2, about 207 feet on a side.

1.11 The analysis of wind turbine performance does not indicate superior efficiency for larger values of rotor radius R. Why in practice are larger wind turbines being adopted? (Hint, the wind speed increases on average with increasing height.)

Exercises to Chapter 2

2.1 Compare the density at the center of the sun to the density of water.

2.2 Find the temperature T needed to bring two alpha particles so close that they "touch" (this might allow them to "fuse" to form a larger particle of mass 8). Each particle has a charge positive $2e$, where $e = 1.6 \times 10^{-19}$ Coulomb. Each particle has 4 atomic mass units, $A = 4$ (the alpha particle is composed of two protons and two neutrons, very strongly bound together by the nuclear force). The useful formula for the radius of a nucleus (like the alpha particle) is $R = R_o A^{1/3}$, where $R_o = 1.44$ fm $= 1.44 \times 10^{-15}$ m. Find the potential energy $U = k_c Q^2/r$, where $k_c = 9 \times 10^9$ (in SI units) is the Coulomb constant. Then find the needed temperature T.

2.3 In the physics of small particles, like electrons and alpha particles, quantum mechanics uses a wavefunction $\psi(x)$ such that $\psi^*\psi$ is the probability density for finding the particle. The particle of mass m can tunnel through a barrier, say of height (barrier energy–energy of particle) φ, and thickness t, with a probability $T \sim \exp(-2\gamma)$, where a simple formula for γ is $(2m\varphi)^{1/2}t/\hbar$, where $\hbar = h/2\pi$ and h is Planck's constant. Find the value of tunnel probability T for an electron mass (9.11×10^{-31} kg), $\varphi = 5$ eV, and $t = 0.1$ nm. (Hint, you need to convert to SI units, so energy must be converted to Joules.)

2.4 Find the de Broglie wavelengths h/p of an electron of energy 1000 eV and a deuteron of energy 57.5 keV.

2.5 In Chapter 2, the power output from the sun is analyzed from the point of view of proton–proton fusion, starting in Equations 2.25–2.28, then finally Equation 2.33 for the power per cubic meter generated. This problem is to see how raising the temperature from 15 million K to 16 million K will change the power output. Recalculate the tunneling probability $T \approx 10^{-8}$ (Equations 2.21–2.28) and the power density (Equation 2.33) for the new temperature $T = 1.6 \times 10^7$. (Consider the effect on the energy E, which affects the parameter r_2, and also on the mean speed that was 0.498×10^6 m/s.)

Exercises to Chapter 3

3.1 Using the equations for the Bohr model of the hydrogen atom, find the speed and orbital frequency of an electron in the $n=2$ orbit. In the $n=2$ orbit, the radius is 4×0.0529 nm and the angular momentum is $h/2\pi$, where $h = 6.6 \times 10^{-34}$ J s. The mass of the electron is $m_e = 9.1 \times 10^{-31}$ kg.

3.2 Using Equations 3.50–3.57 and values in Table 3.2, find a precise value of the Fermi energy in eV in pure Si at 300 K. Why is there a shift from the midgap location, $E = 0.56$ eV?

3.3 Using Equations 3.50–3.57 and values in Table 3.2, including $E_G = 1.12$ eV and $m_{dos} = 0.19$, find for *pure* Si
   a) the densities of electrons and holes at 300 K;
   b) the value of n_i; and
   c) *the resistivity.* (You will need to evaluate the quantities N_C and N_V from Equations 3.53 and 3.55, respectively.)

3.4 Consider an abrupt PN junction in Si assuming the acceptor concentration is 10^{21} m^{-3} and the donor concentration is 10^{23} m^{-3}. Do this approximately by assuming in each case the majority carrier concentration equals the dopant concentration (e.g., $N_e = N_D$), and use results from Problem 3 to find the Fermi energies on N and P sides. The difference of these two Fermi energies is eV_B.

3.5 Estimate, for the PN junction of Problem 4, the *minority carrier* concentrations on the N (p_n) and P (n_p) type sides using numbers from Problems 3 (n_i) and 4 (majority conc. N_e, N_h), via the product rule, using the gap value 1.12 eV.

3.6 Make an estimate of the reverse current density J_o at 300 K for the PN junction described above. Use $J_o = e[D_n n_p/L_n + D_h p_n/L_h]$ using the minority carrier concentrations you got in Problem 5. Assume $D_n = 40$ cm^2/s and $L_n = 140$ μm (for minority electrons in the P-region); and $D_h = 2$ cm^2/s and $L_h = 14$ μm (for minority holes in the N-region).
   a) Find the current density in A/m^2.
   b) By what factor does this current change if the temperature is raised by 10 K to 310 K?

3.7 A PN junction in silicon with bandgap 1.12 eV exhibits a reverse current density $J_o = 1$ pA/m^2 at 300 K. What value do you expect at 273 K for the same junction?

3.8 The Si PN junction of Problem 7 at 300 K is illuminated with 184 W/m^2 at photon energy 2.0 eV. Assume all of the light is absorbed and all of the resulting minority carriers cross the junction with no recombination.
   a) What is the short-circuit current density?
   b) What is the open-circuit voltage at 300 K?
   c) What is the open-circuit voltage at 273 K?

3.9 Explain in words why the resistivity of a pure and perfectly crystalline semiconductor increases with temperature. Invoke the Kronig–Penney model, and recall that the heat capacity per mole of a solid is typically $3R$.

Exercises to Chapter 4

4.1 Estimate the ion current in the device shown in Figure 4.1. Assume that the radius r around the tip center leading to 100% ionization of the deuterium is 600 nm. The assumption is that all D_2 molecules that fall on the front half of this surface (of area $2\pi r^2$) contribute $2e$ to the ion current. The rate of molecules crossing a surface in a dilute gas is $nv/4$ (molecules per unit area per unit time), where n is the number per unit volume of molecules whose rms speed is v.

  a) In a gas of deuterium molecules at temperature $T = 270$ K, show that the rms speed is 1294 m/s. The deuterium molecule has a mass of 4 amu.

  b) The gas is said to be at a pressure of 0.7 Pa at $T = 270$ K. Using the ideal gas law $pV = RT$ (for one mole, corresponding to Avogadro's number of molecules), show that the number density n of molecules is $1.9 \times 10^{20}/\text{m}^3$. On this basis, show that the ion current would be about 44.5 nA (the observed value is 4 nA).

  c) Verify that the mean free path of the D ion exceeds the dimension of the container, so that straight-line trajectories can be assumed. A formula for the mean free path is $\lambda = 1/n\sigma$, where the cross section can be taken as $\sigma = \pi \varrho^2$, and take the radius ϱ of the D_2 molecule as 0.037 nm.

4.2 Adapt the analysis in Chapter 2, Equation 2.21 and following to the case of two deuterons assumed to fuse to form an excited alpha particle (mass 4, charge 2). Assume the kinetic energy of the single deuteron is 115 keV. Find the tunneling probability (Gamow factor) for this reaction. You have to find the energy in the center of mass frame of reference and use the reduced mass of the two particles in the calculation.

4.3 In Chapter 2, the power output from the sun is analyzed from the point of view of proton–proton fusion (Equation 2.21), Equations 2.25–2.28, then finally Equation 2.33 for the power per cubic meter generated, which we will take as 313 W/m^3. This problem is to scale this analysis to find the power density in a D–D reaction Tokamak reactor assuming deuterium density 10^{20} m^{-3}. Assume a factor of 10 increase in T from 15 million K to 150 million K. Scaling must include the $r2$ parameter, setting the T reactivity to 1.0, note that the mass of the deuteron is twice that of the proton, and the temperature, which enters the velocity and also the cross section for the geometric collision. If the Tokamak has a volume 500 m^3 how much power does it release in fusion reactions (assume $Q = 3.5$ MeV for the D–D reaction).

4.4 Following from the Tokamak plasma (Problem 3) assuming deuterium density 10^{20} m^{-3}. $T = 150$ million K. If the Tokamak has a volume 500 (489.5) m^3 the fusion power, assume $Q = 3.5$ MeV for the D–D reaction, is about 84.7 megawatts. Following are questions about this situation:

  a) How much energy U is needed to heat the 500 m^3 deuterium to 1.5×10^8 K, where we know it is 12.93 keV per particle? Express this in kWh and find the dollar value assuming 14 c/kWh.

  b) What is the electron thermal velocity v in the plasma?

c) What is the mean free path λ for scattering of the electrons by the ions? Assume the same parameter r2 (111.3 f at 12 930 eV) determines the collision cross section, so $\lambda = 1/[n\pi(r2)^2]$.
d) Using the answers to (b) and (c), find the collision frequency $f = 1/\tau$ for an electron, $f = v/\lambda$, note that τ the scattering time is $1/f$ and is used to find the mobility. Neglect the small-angle scattering correction $\ln(\Lambda)$ (Equation 4.40) in this problem.
e) Now find the mobility $\mu = e\tau/m$ and the electrical conductivity $\sigma = ne\mu$ (units $m^2/V\,s$). Explain why you can neglect the ions for this calculation.
f) Find the resistivity $\varrho = 1/\sigma$ in ohm m.
g) Express this value as a multiple of the value for Cu metal at room temperature, $\varrho = 2.0 \times 10^{-8}\,\Omega\,m$.
h) Consider a torus of volume $489\,m^3$ (about 500) and this has major radius 6.2 m and minor radius 2 m. Think of this as a cylinder of length $2\pi\,6.2\,m$ and cross-sectional area $\pi 2^2$. Using the formula for the resistance R of a length L with cross section A, $R = \varrho L/A$, find the resistance around this torus, containing the plasma as described.
i) Suppose we want to heat this plasma at power 10 MW, find the needed voltage using $P = V^2/R$.
j) Using the Faraday law, $EMF = -d\Phi/dt$, where Φ is magnetic flux in Webers, consider a linear solenoid superconducting coil as is used in the MRI apparatus, which provides 10 T and has a bore inner diameter of 0.7 m. We can set it to go from 0 T to 10 T in a specified number of seconds. How many seconds will be needed to get the 10 MW of heating? (This solenoid will be perpendicular to our torus, running through the hole. The hole in the torus has radius $6.2\,m - 2\,m = 4.2\,m$, so there is room for the solenoid to fit.
k) How long will the reaction run at 84.7 MW before the deuterium is 50% burned to helium and has to be replaced?

4.5 Find the power density in a T–T reaction Tokamak reactor assuming triton density $10^{20}\,m^{-3}$ at 150 million K. Scaling must include the $r1$ and $r2$ parameters, setting the T reactivity to 1.0, the mass of the triton is three times that of the proton. If the Tokamak has a volume $500\,m^3$ how much power does it release in fusion reactions (assume $Q = 11.3$ MeV for the T–T reaction).

4.6 In a Tokamak with D particles with magnetic field $B = 1$ T, find the Larmor radius of electrons and of deuterons (assume $kT = 12.93$ keV per particle).

4.7 In a Tokamak with D–D fusion leading to T + p (triton mass 3 and proton mass 1) with energy release $Q = 4.04$ MeV, find the kinetic energy of the triton and the kinetic energy of the proton.

4.8 Following Problem 3, $B = 1$ T, find the Larmor radius for the triton and for the proton. Explain in what directions of motion for the reaction products there would be strongest magnetic deflection.

4.9 Find the energy in Joules and the cost in dollar at 14 c/kwH to create a 1 T magnetic field in the volume of the torus, $500\,m^3$. Find the magnetic pressure on the wall in bars (101 kPa). The magnetic energy density is $B^2/2\mu_0$.

4.10 Find the reduced mass for the muonic deuterium atom Dμ. Note that $m_\mu = 206.76\, m_e$, $m_D = 2 \times 1836.2\, m_e$, and $m_{red} = mM/(m+M)$.

4.11 Find the Bohr radius and binding energy for the Dμ atom. (Do this by scaling from the H-atom values.)

4.12 Suppose two D nuclei come together, to create a nucleus of charge $Z = 2$ and mass $4\, m_p$. In this case, what are the new Bohr radius and binding energy for a muon.

4.13 Since the binding energy of the Dμ atom is much larger than that of the deuterium atom De, explain what will happen if a beam of muons is directed through a gas of ordinary De atoms.

4.14 Following from above, explain why in a dense De gas one might expect formation of DDμμ molecules and associated DDμ ions.

Exercises to Chapter 5

5.1 If the bandgap energy of silicon is 1.12 eV, numerically estimate the fraction of the sun's spectrum that will be absorbed (only the photons having energy greater than 1.12 eV). For this numerical estimate, use the measured solar spectrum (above the atmosphere), Figure 1.3. (The sharp dips in this spectrum come from absorption of atoms in the sun's atmosphere, a side effect not characteristic of "black body radiation" predicted by Planck's Law).

5.2 The "tank to wheels" efficiency of the conventional ICE (internal combustion engine) car is 12.6%, including 62.4% loss (from the ~38% efficiency based on the thermodynamic cycle) directly in the engine from friction and heat. Make a simple estimate (you will need to look up some data) on how much loss could be attributed to reciprocating motion of cylinders versus the continuing rotary motion of the Wankel engine (used in the Mazda RX-8 vehicle).

5.3 The area of New York City is 308.9 square miles. If the solar input is 205 W/m^2 and the solar cell efficiency is 15%, find what fraction of this total area represents rooftops if we take as valid the City of New York prediction that 5.85 GW could be supplied by rooftop solar panels.

5.4 The director of the NASA Ames Research Center is quoted as saying that "space based power is five orders of magnitude more expensive than solar cells in the Arizona desert." What reason did he give for his statement?

5.5 Explain why space-based power could be regarded as a concentrating system.

5.6 Using formula 5.3, find what reaction temperature corresponds to a Carnot efficiency of 38%, assuming an exhaust temperature of 300 K.

Exercises to Chapter 6

6.1 In Figure 6.15, explain why the cell thickness is smaller than the thickness of Si cell such as shown in Figure 6.5. Relate this to an aspect of the band structure (see Figure 3.15).

6.2 "Closed space sublimation" (CSS) seems to be a leading low-cost method for making CdTe solar cells. Can you find some references for this?

6.3 In connection with Figure 6.15, explain the use of an anneal with $CdCl_2$. Explain why this is useful, referring to Figure 6.9.

6.4 Figure 6.12 shows a 14.5% efficient CIGS cell. Explain why the process to make this cell, see text, could be considered expensive.

6.5 In Figure 6.5, what is the utility of the structured upper surface, and why is it particularly useful considering the chemical composition of the cell?

6.6 Compare the efficiencies of the two types of organic cells listed in this chapter (including the dye-sensitized titania cell).

6.7 What is the type of solar cell in the large installation at Nellis Air Force Base, shown in Figure 6.6?

6.8 A semiconductor with bandgap 1.5 eV is doped to form an abrupt PN junction.
   a) On the N-side the Fermi level is at 1.4 eV and on the P-side the Fermi level is at 0.1 eV. What is the built-in voltage V_B of the junction, in volts?
   b) If the junction is operated as a solar cell at 300 K, what is the open-circuit voltage in volts if the short-circuit current is 44 times the reverse current?
   c) What is the maximum open-circuit voltage in volts that can be obtained at 300 K from this junction?

6.9 A strongly illuminated Si solar cell exhibits an open-circuit voltage 0.5 volts at 300 K. The cell has a reverse current density of 1 nA/m^2 at 300 K. What short-circuit current density does the cell have under these conditions?

6.10 A windmill produces 50 KW when the wind speed is 13 mph. What do you expect for the power output of the same windmill at wind speed 20 mph? At constant wind speed, how will the power change if the temperature rises?

6.11 Calculate the width W of the depletion region for the junction shown in Figure 6.1, using formulas in Chapter 3. Explain why light absorbed in a wider region can contribute to the current from the cell.

Exercises to Chapter 7

7.1 Explain why the structures in Figure 7.12 are regarded as tandem cells. What are the compositions used to get the three different bandgaps, all based on silicon?

7.2 A tandem solar cell is made of three materials with bandgaps 1.6, 1.0, and 0.6 eV. What is an upper limit to the open-circuit voltage that could be obtained from such a cell?

7.3 If under 1000 W/m^2 illumination the three PN junctions in Problem 2 are known individually to provide short-circuit currents 7.8 mA/cm^2, 12.3 mA/cm^2, and 9.8 mA/cm^2, what is the maximum short-circuit density that the cell can provide in tandem operation?

7.4 In the concentrating photovoltaic system shown in Figure 7.17, the problem of cooling the tandem cells is addressed by water-cooling. What would be the operating temperature of the tandem cells in this system?

7.5 The dye-based concentration system of Figure 7.14 is optimized if the fluorescent emission photon energy matches the bandgap energy of the solar cells mounted at the edges of the glass plates. Make an argument that if the fluoresence spectrum is a sharp peak at the bandgap energy of the receiving solar cells, the efficiency of converting the fluorescent peak can be 100%.

7.6 From Figure 7.1, estimate the highest possible efficiency of a two-junction tandem cell at 1000 suns illumination.

Exercises to Chapter 8

8.1 In a "drifted germanium detector" what charge in Coulombs would flow upon absorption of a 50 keV photon? Take the energy gap for Ge at $T=0$ from Table 3.2.

8.2 In Figure 8.5, write an equation for the thickness t of the ARC, antireflection coating, designed for $\lambda = 560$ nm making use of the index of refraction of the layer, n_{arc}.

8.3 What feature of the band structure in Figure 8.9 indicates that the material is ferromagnetic?

8.4 If the size of a quantum dot is decreased from 6 to 5 nm, how will the wavelength of the emitted light change when the dot is illuminated by high-energy photons?

8.5 Of the solar cell types listed in Table 8.1, which two are shown as images on the front cover of this book? (Hint, see Figure 6.17 and text following Figure 6.6.)

Exercises to Chapter 9

9.1 What is the motivation for the carbon nanostructure shown on the cover of this book?

9.2 Figure 9.1 describes a water-splitting device that requires only light input and also allows electrical power to be extracted. Compare this device with the "artificial leaf," assuming a silicon tandem cell such as shown in Figure 7.12.

9.3 Why is a single junction solar cell inadequate for a water-splitting device?

9.4 For a hydrogen-powered auto, what advantage is there in using hydrogen gas with a fuel cell driving an electric motor over using hydrogen gas directly as fuel in an internal combustion engine?

9.5 In Figure 9.9, nanotube cloth is demonstrated as an incandescent light source. Do you think this would work in the air, and if so why? (Consider the nature of the covalent bonds in graphene.)

9.6 Make an argument that energy storage by pressurizing air in a large cavern should be much cheaper than pumped hydro energy storage, which has been characterized as costing $0.1/Wh of capacity. (You don't have to build a dam, for one.)

Exercises to Chapter 10

10.1 Discuss the rate of growth in renewable energy capacity in connection with the eventual need to enlarge the electric grid. Make an argument that renewable energy installations of a self-sufficient nature can have growth rates larger than in cases where power must be sold to a grid.

10.2 In Figure 10.1, enter a new data point from the statements that solar panels in Oct. 2011 cost $1.20/Wp and that global installed PV capacity at the end of 2010 was 40 GW.

10.3 Explain why, in Equation 3.69 (in Section 10.2), in spite of $k_B T$ as a prefactor, the open-circuit voltage, in fact, decreases with increasing temperature.

10.4 Why would you expect the efficiency of a parabolic trough solar thermal facility to be lower than the central tower type of solar thermal facility? (In a parabolic trough facility, the distance from a hot surface to the turbine can be quite large.)

10.5 Why is the efficiency of a solar thermal facility typically higher than that of a photovoltaic facility?

10.6 Why is a thermal solar facility (Figure 5.4) more adaptable for energy storage than a photovoltaic facility?

10.7 There are approximately 130 million housing units in the United States with 91 million stated as detached single-family homes or trailers. Assume each of the 91 million detached units has installed a rooftop solar water heater of area $10 \, m^2$, as discussed in Section 9.2. Assume the absorbed sunlight at $205 \, W/m^2$ heats water at 100% efficiency that otherwise would be heated with electricity. Find the total average power that would be absorbed, which otherwise would come directly from the power grid. (Answer is 186.5 GW.)

10.8 In connection with Problem 7, compare the efficiency of the direct water heating with an alternative method that would have 10% efficient solar cells of area $10 \, m^2$, with the electrical output being used solely to heat water. (Answer, it is 10 times more efficient.)

Exercises to Chapter 11

11.1 Compose an ordered list of the most urgent problems that will arise in the legacy world once the price of oil is assured to be greater than $300/bbl.

11.2 Make a list of start-up company products/services that you think would succeed in a legacy country with the price of oil greater than $300/bbl.

11.3 What changes in transportation do you expect to see when the price of oil exceeds $300/bbl?

11.4 An area of business opportunity at present is providing and powering cellular telephone transmitters and towers in off-grid areas including the third world, noting that the number of cell phone subscriptions is ~4.6 billion. Make a prioritized list of independent stand-alone power sources suitable for powering remote cell phone transmitters.

11.5 An area of opportunity at present would seem to be a cellular telephone service that will provide the cell phone and the service along with a reliable independent solar power charging unit, one per phone. What renewable power technology would seem suitable for such a start-up company?

11.6 Warren Buffett famously invested in a railroad that serves the State of Wyoming in the United States, which has extensive coal deposits being rapidly extracted. Do a search and list the foreign countries to which the United States exports the most coal. http://205.254.135.24/cneaf/coal/quarterly/html/t7p01p1.html

Total U.S. exports of coal were 60.8 million tons in 2010, up 46.8% over the previous year.

11.7 Assuming all of the 60.8 million tons of coal, quoted in Exercise 11.6, is oxidized to CO_2, how many tons of carbon dioxide will this be? (The masses of C and O are 12 and 16, respectively. For a rough calculation assume the mass of coal is entirely carbon.) Compare this number to the quoted mass of carbon dioxide in the earth's atmosphere, 5.2×10^{18} kg. (One short ton is 907.2 kg.)

Glossary of Abbreviations

AM 0	Solar spectrum observed above earth's atmosphere
AM 1.5G	Average solar spectrum at ground, assuming typical light path
AMU (u)	Atomic Mass unit, (1/12) of mass of ^{12}C: $u = 1.661 \times 10^{-27}$ kg
ARC	Antireflection coating
Bar	One atmosphere pressure 101 kPa
bbl	Barrel unit of volume $0.159\,m^3 = 42$ U.S. gallons
BSF	Back surface field, a stratagem in solar cells to reduce recombination at the back surface of the cell
BTU	British thermal unit 1054 J
CAGR	Compound annual growth rate
CBD	Chemical bath deposition
CCS	Carbon capture and storage (sequestration)
CIGS	Copper indium gallium selenide, a type of solar cell
CMOS	Complementary metal oxide semiconductor (computer logic)
CSP	Concentrated solar power, solar thermal electric
CSS	Closed space sublimation
CVD	Chemical vapor deposition
D	Debye, unit of electric dipole moment, 3.3×10^{-30} C m
D_{es}	Earth–sun distance 1.496×10^8 km
DOS	Density of states, per unit energy per unit volume, for electrons
E_F	Fermi energy highest filled level in an electron system
E_g	Energy gap of a semiconductor, typically in eV
EOT	Equivalent oxide thickness
ESR	Electron spin resonance
EV	Electric vehicle
eV	Electron volt 1.6×10^{-19} J
fcc	Face centered cubic
Fm	Femtometer 10^{-15} m, size scale of the atomic nucleus
GEO	Geosynchronous earth orbit
GW	Gigawatt 10^9 watts
HCCI	Homogeneous charge compression ignition, allows gasoline ICE to operate at higher compression ratio

Nanophysics of Solar and Renewable Energy, First Edition. Edward L. Wolf.
© 2012 Wiley-VCH Verlag GmbH & Co. KGaA. Published 2012 by Wiley-VCH Verlag GmbH & Co. KGaA.

Hcp	Hexagonal close packed
HEV	Hybrid electric vehicle
HOMO	Highest occupied molecular orbital
HP	Horsepower, unit of power 745.7 watts
HTS	High-temperature superconductor
HVDC	High-voltage direct current (transmission line)
ICE	Internal combustion engine
IPCE	Incident photon conversion efficiency
IR	Infrared
ISS	International space station
ITO	Indium tin oxide, conductive glass
kWh	Kilowatt hour $= 3.6 \times 10^6$ J
LEO	Low earth orbit
LNG	Liquid natural gas
LUMO	Lowest unoccupied molecular orbital
MBE	Molecular beam epitaxy: deposits atomically perfect layers
M_e	Mass of the earth 5.97×10^{24} kg
meV	Millielectron volts
MeV	Million electron volts
MW	Million watts, megawatt
MWNT	Multiwall carbon nanotube
NiMH	Nickel–metal hydride, a type of battery used in the Prius hybrid auto
Nm	Nanometer 10^{-9} m
NW	Nanowire
OLED	Organic light emitting diode
OMCVD	Organometallic CVD
PANI	Polyaniline
PC	Photonic crystal, personal computer
Pc	Phthalocyanine
PECVD	Plasma enhanced CVD
PHEV	Plug-in hybrid electric vehicle
PLiON	Form of Li-ion battery
PMMA	Polymethylmethacrylate, used as a photoresist in patterning
PN junction	Junction of P (positively doped), N (negatively doped) semiconductors.
PPV	Polyphenylene vinylene
PV	Photovoltaic cell or module. Solar cell.
PW	Petawatt $= 10^{15}$ watts
QUAD	Quadrillion BTU $= 10^{15}$ BTU $= 1.054 \times 10^{18}$ J
QD	Quantum dot, a three-dimensionally small object, "artificial atom"
R_E	Radius of the earth $= 6173$ km
RRR	Residual resistance ratio, measure of metal purity
R_s	Radius of sun $= 0.696 \times 10^6$ km
SAR	Synthetic aperture radar
SEM	Scanning electron microscope

SWNT	Single-wall carbon nanotube	
TEM	Transmission electron microscope	
TPa	TeraPascal, 10^{12} N/m^2 possible value of Young's modulus	
TPES	Total primary energy supply approximately 15 TW, see Figure 1.1	
TCO	Transparent conductive oxide	
TW	Terawatt, 10^{12} watts	
u (AMU)	Atomic mass unit, 1/12 of mass of carbon ^{12}C = 1.661×10^{-27} kg	
UV	Ultraviolet	
WEC	Wave energy conversion device	

Some Useful Constants

Avogadro's number	N_A	6.022×10^{23} particles/mol
Boltzmann's constant	k_B	1.381×10^{-23} J/K
Ideal gas constant	$R = N_A k_B$	8.315 J/(mol K)
Fundamental charge	e	1.602×10^{-19} C
Mass of electron	m_e	9.109×10^{-31} kg (= 511 keV/c^2)
Mass of proton	m_p	1.672×10^{-27} kg (= 938.3 MeV/c^2)
Planck's constant	h	6.63×10^{-34} Js (= 4.136×10^{-15} eV s)
Planck's constant	$\hbar = h/2\pi$	1.055×10^{-34} Js (= 6.58×10^{-16} eV s)
Bohr magneton	$\mu_B = e\hbar/2m_e$	9.274×10^{-24} J/T (= 5.79×10^{-5} eV/T)
Coulomb constant	$k_C = 1/(4\pi\varepsilon_o)$	8.988×10^9 Nm^2/C^2
Permeability of space	μ_o	$4\pi \times 10^{-7}$ N/A^2
Speed of light	c	2.998×10^8 m/s (= 0.2998 mm/ps)
Photon energy	$hc/\lambda = hf$	1240 (eV nm)/nm
Hydrogen atom binding energy	$k_C e^2/2a_o$	13.6 eV
Bohr radius	a_o	0.0529 nm
Electron volt	eV	1.602×10^{-19} J (= 23.06 k Cal/mol)
Stefan–Boltzmann constant	$\sigma_{SB} = 2\pi^5 k_B^4/(15\, h^3 c^2)$	5.67×10^{-8} W/m^{2}K^4
Gravitational constant	G	6.67×10^{-11} J m^2/kg^2

References

1 McElroy, M.B., Lu, S., Nielsen, C.P., and Wang, Y. (2009) Potential for wind-generated electricity in China. *Science*, **325**, 1378.
2 Bradsher, K. (2011) China's utilities cut energy production, defying Beijing, *New York Times*, May 24.
3 Richter, B. (2010) *Beyond Smoke and Mirrors: Climate Change and Energy in the 21st Century*, Cambridge University Press.
4 Thuillier, G., Herse, M., Labs, D., Foujols, T., Peetermans, W., Gillotay, D., Simon, P., and Mandel, H. (2003) *Solar Phys.*, **214**, 1.
5 Lewis, N. and Crabtree, G. (eds) (2005) *Basic Research Needs for Solar Energy Utilization, Report of Workshop April 18–21, 2005*, U. S. Department of Energy, Office of Basic Energy Sciences. See also, National Research Council and National Academy of Engineering (2004) *The Hydrogen Economy*, The National Academies Press, Washington. On p. 30 the total installed electric capacity in the U.S. in 2004 is stated as 786 GW.
6 Cost of solar cells and of solar panels is expressed in $/Wp (dollars per peak Watt) that references the power the device will produce under 1000 W/m^2. The cost of electricity is commonly quoted in $/kWh and in the United States. This is often near 10 c/kWh. It is thought that a facility based on panels costing $ 1/Wp can market electricity at about $0.06/kWh.
7 Turcotte, D.L. and Schubert, G. (2002) *Geodynamics*, Cambridge Univ. Press.
8 Atzeni, S. and Meyer-Ter-Vehn, J. (2004) *The Physics of Inertial Fusion*, Oxford University Press, Oxford.
9 Schlattl, H. (2001) *Phys. Rev.*, **D64**, 013009.
10 Haxton, W.C. (1995) *Annu. Rev. Astrophys.*, **33**, 495.
11 Bethe, H.A. and Critchfield, C.L. (1938) *Phys. Rev.*, **54**, 248.
12 Gamow, G. (1928) *Z. Phys.*, **51**, 204.
13 Scruggs, J. and Jacob, P. (2009) *Science*, **323**, 1176.
14 MacKay, D.J.C. (2009) *Sustainable Energy: Without the Hot Air*, UIT, Cambridge, England, p. 74.
15 MacKay, D.J.C., op cit, p. 82, Figure 14.3.
16 MacKay, D.J.C., op. cit., p. 81.
17 Turcotte, D.L. and Schubert, G. (2002) *Geodynamics*, Cambridge Univ. Press.
18 Bahcall, J.N., Pinsonneault, M.H., and Basu, S. (2001) *Astrophys. J.*, **555**, 990. See Table 5.
19 Zirker, J.B. (2002) *Journey from the Center of the Sun*, Princeton University Press, p. 11.
20 Phillips, K.J.H. (1995) *Guide to the Sun*, Cambridge University Press, p. 47.
21 Griffiths, D.J. (2005) *Introduction to Quantum Mechanics*, 2nd edn, Pearson Education, p. 141.
22 Griffiths, D.J., op. cit., p. 323.
23 Angulo, C., Arnould, M., Rayet, M., Descouvemont, P., Baye, D., and Leclercq-Villain, C. et al. (1999). *Nucl. Phys*, **A656**, 3–187.
24 Bethe, H.A. and Critchfield, C.L. (1938) *Phys. Rev.*, **54**, 248.
25 Rolfs, E. and Rodney, W.S. (1988) "Cauldrons in the Cosmos: Nuclear

Astrophysics" (Univ. of Chicago Press, Chicago, 1988) Figure 6.1.
26 Clayton, D.D. (1968) *Principles of Stellar Evolution and Nucleosynthesis*, University of Chicago Press (see Problem 5.4), p. 369.
27 Clayton, D.D. (1968) *Principles of Stellar Evolution and Nucleosynthesis*, University of Chicago, Table 6.6 on p. 483.
28 Venkatramani, N. (2002) Industrial plasma torches and applications. *Curr. Sci.*, **83**, 254. Table 5, p. 259.
29 Park, D. (1992) *Introduction to the Quantum Theory*, 3d edn, McGraw-Hill.
30 Anderson, P.A. (1959). *Phys. Rev.*, **115**, 553.
31 Chandrasekhar, S. (1983), "On Stars, their evolution and their stability." Nobel Lecture 8 Dec. 1983, see Eq. 27. This lecture gives a clear discussion of conditions in the core of the sun.
32 Bahcall, J.N., Pinsonneault, M.H., and Basu, S. (2001) *Astrophys. J.*, **555**, 990, See Table 5.
33 Pilar, F.L. (1990) *Elementary Quantum Chemistry*, 2nd edn, McGraw-Hill, Dover, p. 125.
34 A simple description of the universal van der Waals attractive energy, $U = -k_c^2 c_{vdw}/r^6$ between two atoms at spacing r is given in "Nanophysics and Nanotechnology, 2nd edn," by Edward L. Wolf (Wiley-VCH, 2006) pp. 86–90. Here, c_{vdw} is proportional to the product $Z_1 Z_2$ of the number of electron orbitals Z in each atom, and to a mean ionization energy, and k_c is the Coulomb constant.
35 Kittel, C. (1986) *Introduction to Solid State Physics*, 6th edn, John-Wiley & Sons, Inc., p. 55.
36 Kittel, C. (1986) *Introduction to Solid State Physics*, 6th edn, John-Wiley & Sons, Inc., p. 235.
37 Mott, N.F. (1973) Metal–insulator transitions. *Contemp. Phys.*, **14**, 401. Mott's numerical estimate is for an $n = 1$ nodeless radial wavefunction, suitable for hydrogen and hydrogenic impurities in semiconductors, although it is clear that his argument would give even faster delocalization for states such as the 6s state mentioned in the text.
38 Lang, N.D. and Kohn, W. (1971) *Phys. Rev. B*, **3**, 1215.
39 Pauling, L. and Wilson, E.B. (1935) *Introduction to Quantum Mechanics*, McGraw-Hill, pp. 326–336.
40 Griffiths, D.J. (2005) *Quantum Mechanics*, 2nd edn, Pearson Education.
41 Prince, M.B. (1955) *J. Appl. Phys.*, **26**, 534.
42 Miyamoto, K. (2004) *Plasma Physics and Controlled Nuclear Fusion*, Springer.
43 Atzeni, S. and Meyer-Ter-Vehn, J. (2004) *The Physics of Inertial Fusion*, Oxford University Press, Oxford.
44 MacKay, David J.C. (2009) *Sustainable Energy: Without the Hot Air*, UIT, Cambridge, England.
45 Naranjo, B., Gimzewski, J.K., and Putter, S. (2005) *Nature*, **434**, 52.
46 Ruggiero, A.G. (1992) Nuclear fusion of protons with boron. Conference "Prospects for Heavy Ion Inertial Fusion," Crete, Sept. 1, 1992, Brookhaven National Lab., Accelerator Physics Technical Note No. 48.
47 Jackson, J.D. (1957) *Phys. Rev.*, **106**, 330.
48 Rolfs, C.E. and Rodney, W.S. (1988) *Cauldrons In the Cosmos*, Univ. Chicago Press.
49 Oppenheimer, J.R. (1928) *Phys. Rev.*, **31**, 66.
50 Li, X.Z., Tian, J., Mei, M.Y., and Li, C.X. (2000) *Phys. Rev. C*, **61**, 024610.
51 Huba, J.D. (2006) NRL Plasma Formulary, Beam Physics Branch, Plasma Physics Division, Naval Research Laboratory, Washington, DC 20375, p. 57.
52 Miyamoto, K. (2004) *Plasma Physics and Controlled Nuclear Fusion*, Springer, Berlin, Figure 6.3, p. 67.
53 Marti, A. and Luque, A. (eds) (2004) *Next Generation Photovoltaics: High Efficiency Through Full Spectrum Utilization*, Institute of Physics Publishing, Bristol, Figure 13.7, p. 307.
54 MacKay, David J.C. (2009) *Sustainable Energy: Without the Hot Air*, UIT, Cambridge, England, p. 57.
55 Green, Martin A. (2003) *Third Generation Photovoltaics: Advanced Solar Energy Conversion*, Springer, Berlin, (Fig. 1.2).
56 Woody, T. (Sept. 9, 2009) U.S. company and China plan solar project, *New York Times*.
57 Navarro, M. (June 16, 2011) Mapping sun's potential to power New York, *New York Times*.

58. Glaser, P.R. (1968) *Science*, **162**, 857.
59. Potter, Seth D. *et al.* (1999) *Architecture Options for Space Solar Power*, Boeing Co., Downey, CA, http://www.ssi.org/Potter_SSP_99_SSI.pdf. Accessed 6/13/2012.
60. Feingold, H. *et al.* (1997). NASA Space Solar Power: A Fresh Look at the Feasibility of Generating Solar Power in Space for Use on Earth, *NASA Report Number SARC-97/1005*.
61. Morena, L.C., James, K.V., and Beck, J. (2004) An introduction to the RADARSAT-2 mission. *Can. J. Remote Sensing*, **30**, 221, Fig. 3.
62. Boain, Ronald J. and (Jet Propulsion Lab, Cal. Inst. Tech.) (2004) ABC's of Sun-synchronous orbit mission design. Proc. AAS/AIAA Space Flight Mechanics Conference, Maui, Hawaii, 9 Feb. 2004.
63. Shockley, W. and Queisser, H.J. (1961) *J. Appl. Phys.*, **32**, 510.
64. Marti, A. and Luque, A., Eds. (2004) op. cit. Figure 3.5, p. 58.
65. Sze, S. M. (1969) *Physics of Semiconductor Devices*, Wiley-Interscience 1969, Figure 27, p. 54.
66. Wurfel, P. (2009) *Physics of Solar Cells*, 2nd edn, Wiley-VCH Verlag GmbH, Weinheim, Figure 7.6, p. 174.
67. Zhao, J., Wang, A., Green, M., and Ferrazza, F. (1998) *Appl. Phys. Lett.*, **73**, 1991.
68. Zhao, J. *et al.* (1995) *Appl. Phys. Lett.*, **66**, 3636.
69. Alferov, Z.I. and Rumyantsev, V.D. (2003) Chapter 2: Trends in the development of solar photovoltaics, *Next Generation Photovoltaics* (eds A. Marti and A. Luque), Institute of Physics, Bristol and Philadelphia, Figure 2.5, p. 27.
70. Lewis, N. and Crabtree, G. (2005) *Basic Research Needs for Solar Energy Utilization, Report of Workshop, April 18–21*, U.S. Department of Energy, Office of Basic Energy Sciences.
71. Seeger, K. (1982) *Semiconductor Physics*, 2nd edn, Springer, New York, pp. 148–152.
72. Shockley, W. and Queisser, H. (1961) *J. Appl. Phys.*, **32**, 510.
73. (a) Green, Martin A., (2003) *Third Generation Photovoltaics*, Springer, Berlin, (Equations 4.9–10, pp. 38–39); (b) Kondo, M. and Matsuda, A. (2010) Chapter 8: Low temperature fabrication of nanocrystalline-silicon solar cells, in *Thin-Film Solar Cells* (ed. Y. Hamakawa), Springer, Berlin, Figure 8.1, p. 140.
74. Global Climate and Energy Project (GCEP) (2006) *An Assessment of Solar Energy Conversion Technologies and Research Opportunities*, Global Climate and Energy Project, Stanford University.
75. Kapur, F V., Bansal, A., Le, P., Asensio, O., and Shigeoka, N. (2003) Proceedings of the Conference on Photovoltaic Energy Conversion, vol. 1, p. 465.
76. Bremaud, D., Rudmann, D., Bilger, G., Zogg, H., and Tiwari, A. (2005) Record of 31st IEEE Photovoltaic Specialists Conference, p. 658.
77. Negami, T., Aoyagi, T., Satoh, T., and Shimakawa, S. (2002) Record of 29th IEEE Photovoltaic Specialists Conference, p. 658.
78. Negami, T., Aoyagi, T., Satoh, T., and Shimakawa, S. (2002) Record of 29th IEEE Photovoltaic Specialists Conference, p. 658.
79. Ramanathan, K., Noufi, R., To, B., Toung, D., Bhattarcharya, R., Contreras, M., Dhere, B., and Teeter, G. (2006) Proceedings of the 4th IEEE Worldwide Conference on Photovoltaic Energy Conversion, vol. 1, p. 380.
80. Woody, T. (2009) U.S. company and China plan solar project, *New York Times*, Sept. 9. See also Yergin, D. 2011 *The Quest*, Penguin, New York, p. 581.
81. Deb, S.K. (2010) in *Thin-Film Solar Cells* (ed. Y. Hamakawa), Springer, Berlin, Figure 2.5, p. 25.
82. Graetzel, M. (2001) *Nature*, **414**, 138.
83. Junghanel, M. (2007) Novel aqueous electrolyte films for hole-conduction in dye sensitized solar cells and development of an electron transport model. Dissertation, Freie University of Berlin.
84. Nazeeruddin, M.K. *et al.* (2001) *J. Am. Chem. Soc.*, **123**, 1613.
85. Nazeeruddin, M.K. *et al.* (2007) *J. Photochem. Photobiol. A.*, **185**, 331.
86. Rostalski, J. and Meissner, D. (2000) *Sol. Energ. Mater. Sol. Cells*, **61**, 87.
87. Algora, C. (2004) Chapter 6, in *Next Generation Photovoltaics: High Efficiency Through Full Spectrum Utilization* (eds A.

Marti and A. Luque), Institute of Physics Publishing, Bristol.
88 King, R.R. et al. (2009) 24th European Photovoltaic Solar Energy Conference, Hamburg, Germany, Sep. 21–25.
89 King, R.R. et al. (2002) 29th IEEE Photovoltaic Specialists Conference, New Orleans, LA, May 20–24.
90 Karam, N.H., Sherif, R.A., and King, R.R. (2004) Chapter 10: Multijunction concentrator solar cells: an enabler for low-cost photovoltaic systems, in *Next Generation Photovoltaics: High Efficiency Through Full Spectrum Utilization* (eds A. Marti and A. Luque), Institute of Physics Publishing.
91 The Economist (2008) Another silicon valley? The rise of solar energy in one form or another, *The Economist*, June 21.
92 Kim, J.Y., Lee, K., Coates, N., Moses, D, Nguyen, T., Dante, M., and Heeger, A. (2007) *Science*, **317**, 222.
93 Banerjee, A., DeMaggio, G., Lord, K., Yan, B., Liu, F., Xu, X., Beernink, K., Pietka, G., Worrel, C., Dotter, B., Yang, J., Guha, S., and (United Solar Ovonics, LLC) (2009) Advances in cell and module efficiency of a-Si:H based triple-junction solar cells made using roll-to-roll deposition. IEEE Conference Proceedings 978-1-4244-2950-9/09, p. 000116.
94 Carlson, David E. (2010) The Status and Future of the Photovoltaics Industry http://www.aps.org/units/gera/meetings/march10/upload/CarlsonAPS3-14-10.pdf. Accessed 6/13/2012.
95 Currie, J.J., Maple, J.K., Heidel, G.D., Goffri, S., and Baldo, M.A. (2008) *Science*, **321**, 226.
96 Barnett, A., Honsberg, C., Kirkpatrick, D., Kurtz, S., Moore, D., Salzman, D., Schwartz, R., Gray, J., Bowden, S., Goossen, K., Haney, M., Aiken, D., Wankass, M., and Emery, K. (2006) Record of the 4th IEEE Worldwide Conference on Photovoltaic Energy Conversion, vol. 2, p. 2500.
97 University of Delaware (2007) UD-led team sets solar cell record, joins Dupont on $100 million project, University of Delaware News Release, July 23.
98 Sierra Magazine, Sept/Oct 2011. Innovate/solar flotilla.
99 Woody, T. (April 19 2011) Solar on the water, *New York Times*.
100 Luque, A. and Marti, A. (1997) *Phys. Rev. Lett.*, **78**, 5014.
101 Marti, A., Cuadre, L., and Luque, A. (2004) Chapter 7: Intermediate band solar cells, in *Next Generation Photovoltaics* (eds A. Marti and A. Luque), Institute of Physics, Bristol.
102 Honsberg, C., Barnett, A., and Kirkpatrick, D. (2006) Record of the 4th IEEE Worldwide Conference on Photovoltaic Energy Conversion, vol. 2, p. 2565.
103 Marti, A., Cuadre, L., and Luque, A. (2004) Chapter 7: Intermediate band solar cells, in *Next Generation Photovoltaics* (eds A. Marti and A. Luque), Institute of Physics, Bristol.
104 Marti, A., Antolin., E., Stanley, C., Farmer, C., Lopez, N., Diaz, P., Canovas, E., Linares, P., and Luque, A. (2006) *Phys. Rev. Lett.*, **97**, 247701.
105 Lewis, N. and Crabtree, G. (eds) (2005) Basic Research Needs for Solar Energy Utilization: Report of Workshop, April 18–21, 2005, U.S. Department of Energy, Office of Basic Energy Sciences.
106 Schaller, R. and Klimov, V. (2004) *Phys. Rev. Lett.*, **92**, 18660.
107 Olsson, P., Domain, C., and Guillemoles, J.F. (2009) *Phys. Rev. Lett.*, **102**, 227204.
108 Carlson, David E. (2010) op. cit.
109 Turner, J.A. (2004) *Science*, **305**, 972.
110 Fletcher, E.A. and Moen, R.L. (1977) *Science*, **197**, 1050; Weimer, A.W. (2005) Solar thermochemical splitting of water. Paper presented at Global Climate and Energy Project (Stanford University) slides 8 and 20, February 24, 2006 (The ~18% efficient central receiver solar thermal electric facility is in Seville, Spain, operating since 2005. See Figure 5.4 and following text, where its efficiency was estimated at 14.7%).
111 Khaselev, O. and Turner, J. (1998) *Science*, **280**, 425.
112 Graetzel, M. (2001) *Nature*, **414**, 338.
113 Service, R.F. (2011) Artificial leaf turns sunlight into a cheap energy source. *Science*, **332**, 25.

114 Packler, M. (June 17, 2008) Latest Honda runs on hydrogen, not petroleum, *The New York Times*.

115 The Economist (2007) Compressed air might help to make wind power more reliable, *The Economist*, July 25.

116 Innovative hydrogen liquefaction cycle, Gas Equipment Engineering Corp., in DOE 2007 Merit Review, May 15, 2007.

117 Patchkovskii, S., Tse, J., Yurchenko, S., Zhechkov, L, Heine, T., and Seifert, G. (2005) *Proc. Natl. Acad. Sci. U.S.A.*, **102**, 10439.

118 (a) Durgun, E., Ciraci, S., Zhou, W., and Yildirim, T. (2006) *Phys. Rev. Lett.*, **97**, 22612; (b) Sigal, A., Rojas, M.I., and Leiva, E.P.M. (2011) *Phys. Rev. Lett.*, **107**, 158701.

119 Lee, H., Choi, W., and Ihm, J. (2006) *Phys. Rev. Lett.*, **97**, 056104.

120 Zhang, M., Fang, S., Zakhidov, A., Lee, S., Aliev, A., Williams, C., Atkinson, K., and Baughman, R. (2005) *Science*, **309**, 1215.

121 U.S. Geological Survey, Mineral Commodity Summaries, January 2005.

122 Lewis, N. and Crabtree, G. (eds) (2005) Basic Research Needs for Solar Energy Utilization, Report of Workshop, April 18–21, 2005, U.S. Department of Energy, Office of Basic Energy Sciences.

123 Pinto, M. (2007) Nanomanufacturing technology: Exa-units at nano-dollars, in *Future Trends in Microelectronics: Up the Nano Creek* (eds S. Luryi, J. Xu, and A. Zaslavsky), John Wiley & Sons, Inc., Hoboken, p. 154.

124 Nocera, J. (Sept. 23 2011) The phony Solyndra scandal, *New York Times*.

125 Howell, D. (Feb. 4 2011) Energy department aims to slash solar costs 75%, Investors.com.

126 Osborne, M. (2010) Manufacturing cost per watt at First Solar falls to US$0.76: module faults nit earnings. Available at http://Wpww.pvtech.org/news/print/manufacturing_cost_per_watt_at_first_solar_falls_to_us0.76_cents_module_fau. Accessed 6/13/2012.

127 Wald, Matthew L. (Sept. 7, 2011) 2 more solar companies get U.S. loan backing, *New York Times*.

128 Saito, K., Nishimoto, T., Hayashi, H., Fukae, K., and Ogawa, K. (2010) Chapter 7, in *Thin Film Solar Cells* (ed. Y. Hamakawa), Springer, Berlin, Figure 7.11, p. 132.

129 Markoff, J. (Dec.18, 2007) Start-up sells solar panels at lower-than-usual cost, *New York Times*.

130 Lewis, N. and Crabtree, G. (eds) (2005) Basic Research Needs for Solar Energy Utilization, Report of Workshop, April 18–21, U.S. Department of Energy, Offfice of Basic Energy Sciences.

131 Global Climate and Energy Project (GCEP) (2006) *An Assessment of Solar Energy Conversion Technologies and Research Opportunities*, Global Climate and Energy Project, Stanford University.

132 Richtel, M. and Markoff, J. (February 1, 2008) A green industry takes root in California, *The New York Times*. This article states, "the Nanosolar factory looks more like a newspaper plant than a chip-making factory. The CIGS material is sprayed onto giant rolls of aluminum foil and then cut into pieces the size of solar panels."

133 Grunwald, M. and Eilperin, J. (July 30, 2005) Energy bill raises fears about pollution, fraud, *Washington Post*. The article states that the energy bill of 2005 "includes an estimated $85 billion worth of subsidies and tax breaks for most forms of energy – including oil and gas..."

134 Yergin, D. (2011) *The Quest*, Penguin Press, New York, p. 610.

135 Zakaria, F. (2011) How Will We Fuel the Future, *New York Times*; see also Yergin, D. (ed.) *The Quest*, Penguin Press, New York.

136 The Economist (2007) Where the wind blows, *The Economist*, July 26.

137 BP (2007) Statistical Review of World Energy 2007.

138 Wald, Matthew L. (September 11, 2007) Utility will use batteries to store wind power, *The New York Times*.

139 Mandel, J. (Oct. 15, 2010) DOE promotes pumped hydro as option for renewable power storage, *New York Times*.

140 Tarascon, J. and Armand, M. (2001) *Nature*, **414**, 359.

141 The Economist (2008) The end of the petrolhead: Tomorrow's cars may just plug in, *The Economist*, June 28.

142 Tarascon, J. and Armand, M. (2001) *Nature*, **414**, 359.

143. Ravet, N., Abouimrane, A., and Armand, M. (2003) *Nat. Mater.*, **2**, 702; See also Gorman, J. (Sept. 28 2002) New material charges up lithium-ion battery work-Bigger, cheaper, safer batteries, *Science News*.
144. Witkin, J. (Sept. 23, 2011) With battery from Texas, E.V. bus begins its rounds at Heathrow, *New York Times*.
145. Czisch, G., (2006). Low cost but totally renewable electricity supply for a huge supply area: a European/Trans-European example. Available at http://www2.fz-juelich.de/ief/ief-ste//datapool/steforum/Czisch-Text.pdf. Accessed 6/13/2012.
146. Bradsher, K. (October 20, 2011) Chinese trade case has clear targets, not obvious goals, *The New York Times*;UNEP and Bloomberg New Energy Finance (2011) Global Trends in Renewable Energy Investment 2011. Available at http://fs-unep-centre.org/publications/global-trends-renewable-energy-investment-2011.
147. Yergin, D. (2011) *The Quest*, Penguin Press, New York, p. 587.
148. "Renewables 2011: Global Status Report" http://www.ren21.net/Portals/97/documents/GSR/GSR2011_Master18.pdf; see also "Renewable Energy" http://en.wikipedia.org/wiki/Renewable_energy.
149. Richter, B. (2010) *Beyond Smoke and Mirrors: Climate Change and Energy in the 21st Century*, Cambridge University Press, p. 156.
150. Carlson, David E. (2010) op. cit.
151. Leone, S. (2011) U. N. Secretary-General: Renewables can end energy–poverty, RenewableEnergyWorld.com. Available at http://www.renewableenergyworld.com/rea/news/article/2011/08/u-n-secretary-general-renewables-can-end-energy-poverty?cmpid=WNL-Friday-August26-2011. Accessed 6/13/2012.
152. Yergin, D. (2011) There will be oil, *Wall Street Journal*, Sept. 17. This article states that of the total estimated oil in the earth in 1900, the fraction remaining is 0.58, or 1.4×10^{12} barrels, that is, 42% has been consumed in 110 years.
153. Subramanian, A. (2011) The inevitable superpower: why China's dominance is a sure thing, *Foreign Affairs*, Sept.–Oct., p. 66.
154. Richter, B. (2010) *Beyond Smoke and Mirrors: Climate Change and Energy in the 21st Century*, Cambridge University Press, pp. 76–80.
155. MacFarlane, R. and Woolsey, R.J. (Sept. 20, 2011) How to weaken the power of foreign oil, *The New York Times*.
156. Czisch, G. (2006) op. cit.
157. Zakaria, F. (Sept. 23, 2011) How will we fuel the future?, *New York Times*.

Index

a

absorption 9, 51, 83, 118, 134, 137, 142, 164, 181, 192, 206
– atomic absorption lines 8
– black body 118
– coefficients for Ge, Si, and GaAs 136, 138
– of crystalline Si 166
– direct heat 117
– effects in atmosphere 8, 9
– energy in semiconducting donor polymer 165
– materials with high absorption coefficients 210
– of neutrons 105
– of one-electron atoms 50
– of photons 2, 136, 154, 177, 188
– – by Si, and similar semiconductors 74
– properties of solar cells 170
– ruthenium-based dye 156
– sunlight reaching earth diminished by 116
adsorption
– based hydrogen storage methods 200
– sites for H_2 198
– for storing hydrogen in high density 196
– of two Ti atoms 198
– weak adsorption energy 196
AlP:Cr cell 184
alpha decay 46
alpha particle 10, 11, 46, 105, 106
– decay 32, 37, 45
aluminum foil 125, 148, 193, 211
ammonia synthesis loop 200
amorphous silicon 146, 167, 168, 205, 207
amorphous silicon:H-based solar cells 166–169
– constructed with amorphous layers of Si and Si–Ge alloys 168
– quantum efficiency 169

Amperes law 111
angular momentum 62, 64, 65, 71
– orientations of 63
– peculiarity 64
antireflection coating 133
artificial leaf energy conversion system 226, 229
artificial leaf of Nocera 193, 194
atoms 60, 64
Avogadro's number 59
azimuthal angle 61, 64, 116

b

back surface field (BSF) 134, 142, 158, 206
bandgap 56, 74, 75, 123, 141, 149, 151, 160, 170, 183, 189
– direct 136, 152
– energy 77, 135, 141–143, 180, 181
– indirect 74, 75, 166, 184
– junction 145
– molecular 163
– for oxide 153
– of silicon 141
– single-bandgap devices 147
– wide-bandgap AlGaAs window 142
band insulator 57
batteries
– energy densities 218
– Li-ion batteries 217–220
– Ni–Cd batteries 218
– NiH batteries 127, 218
– NiMH batteries 220–222
– Prius batteries 221
– rechargeable 217
– technology, investment in 226
binding energy 35, 43, 44
biofuels 9, 225
black body radiation 7–9, 113, 118, 144

black body radiator   116
Bohr orbit   181
Bohr radius   67
Bohr's model   51, 52, 62, 63
– of hydrogen atom   49–52
Boltzmann constant   59
bond angle   72, 73
bound states   33–35
Bragg scattering   53
bremsstrahlung   109
BSF. *See* back surface field (BSF)

c

Canadian Space Agency   128
carbon dioxide emissions   229
carbon nanotubes   199
Carnot efficiency   117, 118
carrier multiplication   177–180
μ-catalysis cycle   100
CdTe/CIGS cells, large-scale scenario   209
– cell efficiency   210, 211
– printing press CIGS/CdTe cell production, economics   211, 212
– profitable private investment   212, 213
– projected total capital need, conditions   212, 213
– size of solar field   210, 211
– solar influx   210, 211
CdTe thin-film cells   123, 151, 204, 206, 208, 210
– analysis   211
– modules   204, 207
– panels, production line   209
– producer First Solar, learning curve   206
– resources   204
– technology   204
– types of   152
cell efficiency   139, 140, 142, 145, 172, 182, 184, 185, 205, 210, 211
charge motion in periodic potential   52, 53
China
– adverse effect on climate   216
– auto sales   226, 227
– avoiding use of coal   229
– business opportunity   229
– hydroelectric power   18
– number of cars   226
– solar cell farm   123, 215
– thin-film solar panels   152
– wind turbine manufacture   229
CIGS ($CuIn_{1-x}Ga_xSe_2$) thin-film solar cells   147, 204, 208
– absorber layer   208
– aluminum foil   193

– CIGS PN junction without Cd   150
– current–voltage curve   150
– ink-printed nonvacuum approach   148
– – advantages   149
– – cross-section image   149
– modules   151, 210
– quantum efficiency   151
– thin-film cells   123, 209
climate change   229
coal supply   228
concentrated solar power (CSP)   119, 209
– niche application   172–174
– thermal power   224
conversion efficiency   136
copper indium gallium selenide (CIGS)   205
Coulomb barrier   12, 14, 30, 35, 39, 42, 89, 90, 93, 100
Coulomb constant   61
Coulomb energy   44, 85
Coulomb force   111
Coulomb repulsion   11, 43–45, 47, 65
covalent bond   60
current–voltage diagrams
– tandem cell   191
CVD reactor   199

d

Davisson–Germer experiment   53
DC power transmission lines   222
DD fusion   25
– average energy release   25
– catalysis by mu mesons   101, 102
– reactor   42
DD plasma analysis   104
– to DT plasma as in ITER, adaption   104–110
DD reactions   87
Debye length   110
deuterium
– electric field ionization of   94–96
– fusion demonstration
– – based on field ionization   88–94
– – based on muonic hydrogen   96–101
– – in larger scale plasma reactors   102, 103
deuterium fusion   88
deuteron density   110
deuteron flux   93
deuterons   6, 13, 25, 27, 30, 35, 42, 51, 87–91, 93, 96, 98, 99, 102, 105, 107, 109
– binding energy   35
diatomic molecules   68
diode I–V characteristic   80–84
direct photovoltaic conversion   2, 4
direct solar influx   6
donor/acceptor impurities   73–75

dual-purpose thin-film tandem cell devices   193
dye sensitization   154, 155
– principle of   155
– solar cells   153, 154
– – efficiency   155
– TiO_2 192

e

earth-based long-term energy resources   22
– earth's deuterium, and potential   25
– geothermal energy   24, 25
– lunar ocean tidal motion   22–24
earth-level peak sun's power spectrum   118
electrical conduction   52, 59, 73, 107, 163, 218, 220
electric conversion systems   117
electric grid   226
electric heating   211
electric power   226
– demand in China   227
electric vehicles EV   220
electrolysis of water   187, 188
electrolytic cell aspects   189
electrolyzer   187
electrolyzing water   216
electron barrier potential energy   94
electron diffraction   31
electron exchange   68
electron–hole pairs   188
– in semiconductor   180
electroplating   189
electrostatic attraction   68, 69
energy
– balance   1, 9, 10, 102
– challenges   226
– conservation of   31
– consumed in United States   4
– demand and supply   227
– – in China   226, 227
– – energy miracle   229, 230
– – oil bubble   227–229
– density   13, 115, 188, 217, 220, 230
– expression   3
– miracle   229, 230
– poverty   226
– security   230
– supply and demand   227
– worldwide total energy usage per person   8
energy bands   53, 54, 56, 73
– missing electron behavior   180
– properties of metal   57–60
– for Si and GaAs   74
– structures silicon   74

energy gaps   57, 74, 78, 79, 167, 175, 176, 192
– of semiconductors   74, 75, 192
Environmental Protection Agency   229
European wind power network   216

f

fabrication methods   203–205, 207
Faraday law   107
Fermi energy   68
Fermi levels
– of ohmic contact   189
Fermi velocity   59
ferromagnetic materials, for solar conversion
– electron density of states   183
– photovoltaic efficiencies   184
field ionizing neutron generator   89
filled atomic shells   65
filling factor behavior   136
First Solar CdTe modules   211, 212
Fischer–Tropsch chemical process   228
fossil fuels   122, 213
fuel cell/electric motor combination   195
fuel-cell electric propulsion   230
fusion power density   109
– scaling from that in the Sun   104
– scaling ratio   104
fusion reactions   10–13, 88

g

GaAs-based tandem single-crystal cells   158, 160–162
– cost   161
– current–voltage curve   160
– measured efficiency   160
– measured open-circuit voltage   161
– quantum efficiencies   159
GaAs multijunction tandem solar cell   141, 205
– technology   188
GaAs triple-junction solar cells   167
GaInP/GaInAs/Ge lattice-matched cell   157
gamma ray photon   178
Gamow peak   93
Gamow's tunneling model   35–42
– deuteron formations   41
– Gamow probability for p–p reaction   37
– geometric collision   37
– proton turning point collisions   41
– rate of fusion   40
– transient protodeuteron events   41
– transmission probability   36
– WKB approximation   36
Gamow transmission factor   99, 100
Gamow tunneling probability   95

gasoline 8
gas tungsten arc welding 42
geosynchronous earth orbit (GEO) satellite 128
global
– consumed power 3
– energy pattern 5
– energy usage 2
– natural power sources 5
– power consumption 227
– trends in renewable energy investment 4
global positioning system (GPS) 125, 132
greenhouse effect 9
grid management 215–217

h

Haber–Bosch process 200, 201
heat absorption 117
heat energy 6, 9, 10, 25, 105, 109, 115, 117, 222
heat exchangers 195
heavy water 87
Heliostat Power Satellite device 126
He–Ne gas mixture 196
hybridization effect 71
hydride storage tanks 196
hydroelectric power 2, 5, 6, 8, 10, 18, 214, 217, 221, 223, 228
hydrogen 2, 6, 201
– Bohr's model 49–52
– burning 11
– creation, off-the-shelf equipment 216
– economics 200
– electrolytic hydrogen generation 22
– filling stations 200
– fuel cell status 194, 195
– gaseous 195
– generation, cells for 187
– ionization 94
– liquefaction cycle, single-pass low-pressure 196
– muonic 96
– normal hydrogen electrode (NHE) 191
– photocatalytic dissociation of water into 188–190
– as potential intermediate 201
– reacting with itself in core of sun 41
– storage and transport 187, 195, 198, 200, 215, 226
– – further aspects 200, 201
– – nanometer-scale schematic diagram 197
– surface adsorption 196
hydrogen molecule ion H_2^+ 69
– dependence on spacing 70

– energy difference 69
– energy E of symmetric case 70
– linear combination of allowed solutions 69
– oscillator frequency 71
– Pauling's treatment 70
hydropower 2, 213, 215, 225
hydrostatic pressure 8, 60

i

impact ionization 177–180
– predicted solar cell efficiency as enhanced by 182
– process, as observed in PbSe nanocrystals 183
India
– avoiding use of coal 229
– nuclear reactor 128
– rural 216
– Tata group, role in 226
insulator 57, 73, 74, 137, 149, 150, 168
intermediate band solar cell 172, 175–177
– based on quantum dots 179
– intermediate band quantum dot solar cell device 178
International Space Station (ISS) 126
inverter technology 213
investment
– in battery technology 226
– power investment in China 4
– profitable private investment 212, 213
– total capital need 212
iridium NEXT network 127
iron phosphate $LiFePO_4$ cathode (LFP) 218
ITER-based electric power plant 108
ITER reactor 113
– design for 111
ITER Tokamak reactor 87

k

Kepler's law 130
kinetic energy 11, 12, 15, 24, 29, 33, 50, 66, 81, 83, 85, 87, 105, 106, 179
Kronig–Penney model 55, 56

l

Laguerre polynomial 62
Larmor frequency, for electrons 106
lead–acid automobile battery 217
Li-drifted germanium detector 172, 177
light radiation 199
Li-ion batteries 217–221
– performance of 220
liquid natural gas (LNG) 187, 195
$LiTaO_3$ crystal 90

lithium 113
low-cost tandem technology 163–169

m

magnetic flux 107
magnetic quantum number 63
Maxwell's equations 32
MBE. *See* molecular beam epitaxy (MBE)
metallic bond 66
metallic electrodes 178
metal properties 57
metal–vacuum boundary 68
metastable 46
minority carriers 79, 81–83, 134, 136, 139, 140, 143, 147, 206
missing electron 180
mobile cellular phone 194
mobility 52, 73–75, 82, 107, 151, 177
module cost, learning strategies 205–207
molecular beam epitaxy (MBE) 157, 181
molecular bonding 69
molecules 60
– hybridization effect 71
– organic, as solar concentrators 169–171
monolithic photovoltaic–photoelectrochemical device 188, 190
monolithic PV-hydrogen electrolytic converter 188
Mott's formula 67
multijunction concentrating solar cells 162, 163
– application to pontoon mounting 173
muon-catalyzed fusion (μCF) 102
muonic deuterium molecular D_2^+ ions 88

n

nanophysics 5, 6, 8, 26, 30, 36, 51, 52, 110, 116
Nanosolar producing printed CIGS cells 212
national energy emergency 213
natural gas 1–3, 22, 30, 195, 200, 201, 215, 228
– price 201
neutrino 11–13, 27, 35, 38, 39, 41
– flux 28
neutron
– decay 39
– densities 45
– emission rate 102
– generation, in compact device 91
nickel–cadmium battery 220
NiMH batteries 218, 220–222
Nocera, artificial leaf 193, 194

Nocera photocatalytic artificial leaf devices 202
normal hydrogen electrode (NHE) 191
N^+P junction solar cell 134
nuclear fusion 5, 10, 12, 27, 33, 90, 102, 226, 230
– reactors 5, 230
nuclear power 2
nuclear properties 43–47
nuclear reactor 128

o

oil production 227–229
– rising price 228
open-circuit voltage 205
orbital angular momentum 63
orbital energy 6
organic molecules, as solar concentrators 169, 170
organic solar cell 156. *See also* polymer organic solar cells; tandem polymer organic solar cell; ZnPc/MPP organic solar cell
oscillator probability 99

p

palladium–silver alloys 198
Pauli's exclusion principle 58, 64
periodic potential 52, 57, 75
photocatalytic water splitting 188
photocurrent flows 189
photoelectric effect 134
photon energy 115, 133, 134, 136, 144, 154, 171, 175
photovoltaic array (PV) 187, 209
– PV-based hydrogen energy storage approach 215
photovoltaic cells 5, 6, 52, 80–84, 122, 187, 222
– efficiency–cost trade-offs 123
– generations of 122–124
photovoltaic conversion efficiencies 185
photovoltaic electric power 216
photovoltaic energy 204
photovoltaic modules 203
– manufacturing learning curve 204
photovoltaic solar cells. *See* photovoltaic cells
Planck radiation law 7
plasma 6, 28, 84, 87, 103, 104, 107–113
– domains in temperature–density representation 85
– electrical heating of 103
– radiation transfer from 109

plasma (*Continued*)
– Tokamak DT plasma   88
– torches   42
plug-in electric automobiles   226
PN junction   80–84
poloidal field   109, 111
polyacetylene   198
polycrystalline silicon   205
polymer organic solar cells   155, 156
positron   11, 27, 35, 37, 38, 76
positronium   76
power electric cars   215
printing press, economics of   211, 212
printing solar cells
– thin-film cells, nanoinks   207–209
Prius hybrid car   218
probability
– backscattering   137
– D–D spacings   99
– density   55
– fusion reaction probability   38, 88
– Gamow probability for p–p reaction   37, 41, 68
– oscillator   99
– tunneling   36, 37, 39, 40, 46, 90, 92, 95
proton–proton fusion   11, 87, 104
proton–proton reaction   12, 28, 35, 37, 38, 40, 42, 87, 92
protons
– density   104
– masses   46
– and neutrons, binding   35
– in the Sun's core   28–30
PV. *See* photovoltaic array (PV)
pyroelectric crystal   89

r

Radarsat antenna   129, 131
Rayleigh scattering   116
rechargeable batteries   217
rectenna   126
refraction   30, 117, 137, 170
renewable energy
– beyond solar and wind   225, 226
– intermittency   187
– investments   224
– in legacy world and   226, 227
– plan for providing to Europe   5
renewable power   3, 5, 123, 214, 215, 223
– usage of   223
resonant energy   101
resonant state   100
rooftop solar cells   229
ruthenium-based dye N945   156

s

Schrodinger's equation   6, 28, 30, 31, 35, 40, 54, 57, 60, 61, 65, 71, 73, 98
– applied to hydrogen atom   61
– for motion of particles   30, 31
– solutions, tetrahedral bonding in   71
– time-dependent   32
– time-independent   32, 33
– using more realistic 3D forms for   73
– wave treatment of matter particles   40
secondary solar-driven sources   14
– flow energy   14–18
– hydroelectric power   18, 19
– ocean waves   20
– – estimate of wave energy   20–22
semiconducting polymer solar cell   156
semiconductor–electrolyte interface   190
semiconductor quantum dots
– electron and hole states in   180–182
– energy threshold, requirement   182
– multiple exciton generation   181
semiconductors   6, 52, 57
– band-edge energies of   192
– carrier concentrations in   76–79
– degenerate metallic   79, 80
– dilute magnetic   182, 185
– doping of   184
– energy gaps, and electronic parameters   75
– excitons in   75, 76
– hydrogenic donors, and excitons in   75, 76
– organic   205
– P-type   147
shielding property   84
Shockley–Quiesser analysis   157
Si–Ge alloys   123, 136
silicon absorbs sunlight   193
silicon crystalline cells   136–140
– thin-film solar cells *vs.* crystalline cells   145, 146
silicon solar cells   203
– price   203, 204
single-junction cells   133–136
single-junction limiting conversion efficiency   141–145
smaller scale concentrator technology   162, 163
sodium–sulfur battery storage   221
solar-based energy. *See* solar energy
solar cells   4, 14, 16, 26, 120, 127, 132, 173, 194, 205, 209, 222, 226, 229
– CIGS (CuIn$_{1-x}$Ga$_x$Se$_2$) thin-film solar cells   147
– dye-sensitized   153
– efficiency (*See* cell efficiency)

- energy gaps and band-edge energies   192
- inexpensive thinfilm tandem cell   192
- output voltages   189
- polymer organic   155
- radiation-induced current density of   135
- thin-film solar cells *vs.* crystalline cells   145, 146

solar converters, in space   126
solar energy   2, 8, 22, 26, 121, 163, 172, 193, 210, 212–214, 221, 222, 228
- conversion   2, 115–132
- utilization   5
solar influx   13, 14, 210, 211
solar photovoltaic global production   224
solar power. *See also* solar energy
- growth   223–225
- with photovoltaics, utilization   125
- satellite   128, 130
- space-based   126–132
- *vs.* wind power   214, 215
solar spectrum photons   190
solar thermal electric power   119–122
- central receiver thermal solar power installations   120
- EuroDISH parabolic mirror Stirling engine   121
solar water heating   224
SolFocus concentrating cell design   162
space-based solar power   126–132
space blanket   131
spectral splitting cells   171, 172
spherical geometry   84
spin angular momentum   64, 65
spin degeneracy   64, 73
Sputnik Russian space program era   124
Stefan–Boltzmann law   7
Stirling engine system   121
storage/grid management, importance
- large-scale utilization   215
-- batteries: lead–acid to lithium to sodium sulfur   217, 218
-- lithium batteries   218–220
-- NiMH   220–222
- off-the-shelf equipment for creating hydrogen   216
- PV-based hydrogen energy storage approach   215
sun as an energy source
- earth-level peak sun's power spectrum   118
- energy density of radiation   13, 115, 188
- photon energy   115, 133, 134, 136, 144, 154, 171, 175
- power density per unit wavelength   115
- properties   115

- system efficiency *vs.* collector temperature   119
- total power density at surface   116
sun-synchronous orbit   129
surface dipole barrier   68
surface energy   66
synthetic aperture radar (SAR)   128

t

tandem polymer organic solar cell   158, 164, 165
tandem semiconducting polymer cells   163–166
- Band-Edge energies   165, 166
- energy-level schematic   165
- performance   166
-- *J–V* characteristics   166
- structure   164
tetrahedral bonding
- angle   72
- angular dependences, for p and d wavefunctions   72
- Schrodinger's equation application   71
- in silicon and related semiconductors   71, 72
-- bond angle   72, 73
-- directed/covalent bonds, connection with   72
- 5-valent donor atoms   75
thermalization   110
thermal radiation   9
thermodynamics   117
thermonuclear fusion   102
thin-film cell. *See* CdTe thin-film cells; CIGS ($CuIn_{1-x}Ga_xSe_2$) thin-film solar cells
thin-film solar cell manufacturers   207
three-layer silicon cells   208
Ti deposition   199
titanium-decorated carbon nanotube cloth   199, 200
titanium-decorated polyacetylene   198
titanium-decorated polymers   198
Tokamak reactors   87, 226
Tokamak plasma   85
- evaluation   109
toroidal DT plasma   87
toroidal magnetic field   105
toxicity, of Cd   124
transmission line   5, 57
transparent conductive oxide (TCO)   167
tungsten trioxide blue light cell   193
tunnel diode interconnect   189
tunneling
- model (*See* Gamow's tunneling model)

tunneling (Continued)
- probability   36, 37, 39, 40, 46, 90, 92, 95
- quantum mechanical tunneling   13
- rates   96, 100

u

ultimate efficiency   136
United States
- average power per person   221
- distribution of wind speeds across   16
- energy consumed in   4, 209
- installation of single-crystal solar cells   140
- map of requirement of land   15
- Prius batteries   221
- total average power   211, 212
- wind turbine manufacture   229
uranium alpha decay series   44

v

vibrational width of motion   98
voltage drop   189

w

wall erosion, Tokamak   113
water electrolysis   187, 188
water oxidation energy level   190
water, photocatalytic dissociation
- dual-purpose thin-film tandem cell devices
-- possibility   193
- hydrogen and oxygen   188–193
- mass production tandem cell water-splitting device
-- possibility   191–193
water splitting
- approach   190
- efficiency   189
- hydrogen production   188
- mass production tandem cell water-splitting device   191–193
- photocatalytic   188–193
- vs.tandem cell   190, 191
- zero photocurrent   190
water turbines   5, 18, 24, 225
wavefunctions   54, 58, 62, 63, 65
- anisotropic   63
wave number   53, 54
wave probability   55
wind power   213
- electric   216
- growth   223–225
- renewable energy   225, 226
- vs. solar power   214, 215
wind turbines   214
- manufacture   229
WKB approximation   36, 95

x

X-ray energies   90

z

zero point energy   21, 98
ZnPc/MPP organic solar cell   156